RELIABILITY, MAINTAINABILITY AND RISK

Also by the same author

Reliability Engineering, Pitman, 1972

Maintainability Engineering, Pitman, 1973 (with A. H. Babb)

Statistics Workshop, Technis, 1974, 1991

Achieving Quality Software, Chapman & Hall, 1995

Quality Procedures for Hardware and Software, Elsevier, 1990 (with J. S. Edge)

Functional Safety, A Straightforward Guide to IEC 61508, Butterworth-Heinemann, 2001
 (with K. G. Simpson)

Reliability, Maintainability and Risk

Practical methods for engineers
Sixth Edition

Dr David J Smith
BSc, PhD, CEng, FIEE, FIQA, HonFSaRS, MIGasE

AMSTERDAM BOSTON HEIDELBERG LONDON NEW YORK OXFORD
PARIS SAN DIEGO SAN FRANCISCO SINGAPORE SYDNEY TOKYO

Butterworth-Heinemann
An imprint of Elsevier Science
Linacre House, Jordan Hill, Oxford OX2 8DP
200 Wheeler Road, Burlington, MA 01803

First published by Macmillan Education Ltd 1981
Second edition1985
Third edition 1988
Fourth edition published by Butterworth Heinemann Ltd 1993
Reprinted 1994, 1996
Fifth edition 1997
Reprinted with revisions 1999
Sixth edition 2001
Reprinted 2002, 2003

Copyright © 1993, 1997, 2001, David J. Smith. All rights reserved

The right of David J. Smith to be identified as the author of this work
has been asserted in accordance with the Copyright, Designs and
Patents Act 1988

No part of this publication may be reproduced in any material form (including
photocopying or storing in any medium by electronic means and whether
or not transiently or incidentally to some other use of this publication) without
the written permission of the copyright holder except in accordance with the
provisions of the Copyright, Designs and Patents Act 1988 or under the terms of
a licence issued by the Copyright Licensing Agency Ltd, 90 Tottenham Court Road,
London, England W1T 4LP. Applications for the copyright holder's written
permission to reproduce any part of this publication should be addressed
to the publisher.

Permissions may be sought directly from Elsevier's Science and
Technology Rights Department in Oxford, UK: phone: (+44) (0) 1865 843830;
fax: (+44) (0) 1865 853333; e-mail: permissions@elsevier.co.uk . You may also
complete your request on-line via the Elsevier Science homepage
(http://www.elsevier.com), by selecting 'Customer Support' and then 'Obtaining
Permissions'.

British Library Cataloguing in Publication Data
Smith, David J. (David John), 1943 June 22-
    Reliability, maintainability and risk. - 6th ed
    1 Reliability (Engineering)   2 Risk assessment
    I Title
    620'.00452

Library of Congress Cataloguing in Publication Data
Smith, David John, 1943-
    Reliability, maintainability and risk: practical methods for
    engineers/David J. Smith. - 6th ed
    p. cm
    Includes bibliographical references and index.
    ISBN 0 7506 5168 7
    1 Reliability (Engineering)   2 Maintainability (Engineering)
    3 Engineering design   I Title.
    TA169.S64 2001
    620'.00452-dc21   00-049380

ISBN 0 7506 5168 7

For information on all Butterworth-Heinemann publications
visit our website at www.bh.com

Printed and bound in Great Britain by Martins the Printers

Contents

Preface

After three editions *Reliability, Maintainability in Perspective* became *Reliability, Maintainability and Risk* and has now, after just 20 years, reached its 6th edition. In such a fast moving subject, the time has come, yet again, to expand and update the material particularly with the results of my recent studies into common cause failure and into the correlation between predicted and achieved field reliability.

The techniques which are explained apply to both reliability and safety engineering and are also applied to optimizing maintenance strategies. The collection of techniques concerned with reliability, availability, maintainability and safety are often referred to as RAMS.

A single defect can easily cost £100 in diagnosis and repair if it is detected early in production whereas the same defect in the field may well cost £1000 to rectify. If it transpires that the failure is a design fault then the cost of redesign, documentation and retest may well be in tens or even hundreds of thousands of pounds. This book emphasizes the importance of using reliability techniques to discover and remove potential failures early in the design cycle. Compared with such losses the cost of these activities is easily justified.

It is the combination of reliability and maintainability which dictates the proportion of time that any item is available for use or, for that matter, is operating in a safe state. The key parameters are failure rate and down time, both of which determine the failure costs. As a result, techniques for optimizing maintenance intervals and spares holdings have become popular since they lead to major cost savings.

'RAMS' clauses in contracts, and in invitations to tender, are now commonplace. In defence, telecommunications, oil and gas, and aerospace these requirements have been specified for many years. More recently the transport, medical and consumer industries have followed suit. Furthermore, recent legislation in the liability and safety areas provides further motivation for this type of assessment. Much of the activity in this area is the result of European standards and these are described where relevant.

Software tools have been in use for RAMS assessments for many years and only the simplest of calculations are performed manually. This sixth edition mentions a number of such packages. Not only are computers of use in carrying out reliability analysis but are, themselves, the subject of concern. The application of programmable devices in control equipment, and in particular safety-related equipment, has widened dramatically since the mid-1980s. The reliability/quality of the software and the ways in which it could cause failures and hazards is of considerable interest. Chapters 17 and 22 cover this area.

Quantifying the predicted RAMS, although important in pinpointing areas for redesign, does not of itself create more reliable, safer or more easily repaired equipment. Too often, the author has to discourage efforts to refine the 'accuracy' of a reliability prediction when an order of magnitude assessment would have been adequate. In any engineering discipline the ability to recognize the degree of accuracy required is of the essence. It happens that RAMS parameters are of wide tolerance and thus judgements must be made on the basis of one- or, at best, two-figure accuracy. Benefit is only obtained from the judgement and subsequent follow-up action, not from refining the calculation.

A feature of the last four editions has been the data ranges in Appendices 3 and 4. These were current for the fourth edition but the full 'up to date' database is available in FARADIP.THREE (see last 4 pages of the book).

DJS

Acknowledgements

I would particularly like to thank the following friends and colleagues for their help and encouragement:

Peter Joyce for his considerable help with the section on Markov modelling;
'Sam' Samuel for his very thorough comments and assistance on a number of chapters.

I would also like to thank:

The British Standards Institution for permission to reproduce the lightning map of the UK from BS 6651;
The Institution of Gas Engineers for permission to make use of examples from their guidance document (SR/24, Risk Assessment Techniques).
ITT Europe for permission to reproduce their failure report form and the US Department of Defense for permission to quote from MIL Handbooks.

Part One
Understanding Reliability
Parameters and Costs

Part One

1 The history of reliability and safety technology

Safety/Reliability engineering has not developed as a unified discipline, but has grown out of the integration of a number of activities which were previously the province of the engineer.

Since no human activity can enjoy zero risk, and no equipment a zero rate of failure, there has grown a safety technology for optimizing risk. This attempts to balance the risk against the benefits of the activities and the costs of further risk reduction.

Similarly, reliability engineering, beginning in the design phase, seeks to select the design compromise which balances the cost of failure reduction against the value of the enhancement.

The abbreviation RAMS is frequently used for ease of reference to reliability, availability, maintainability and safety-integrity.

1.1 FAILURE DATA

Throughout the history of engineering, reliability improvement (also called reliability growth) arising as a natural consequence of the analysis of failure has long been a central feature of development. This 'test and correct' principle had been practised long before the development of formal procedures for data collection and analysis because failure is usually self-evident and thus leads inevitably to design modifications.

The design of safety-related systems (for example, railway signalling) has evolved partly in response to the emergence of new technologies but largely as a result of lessons learnt from failures. The application of technology to hazardous areas requires the formal application of this feedback principle in order to maximize the rate of reliability improvement. Nevertheless, all engineered products will exhibit some degree of reliability growth, as mentioned above, even without formal improvement programmes.

Nineteenth- and early twentieth-century designs were less severely constrained by the cost and schedule pressures of today. Thus, in many cases, high levels of reliability were achieved as a result of over-design. The need for quantified reliability-assessment techniques during design and development was not therefore identified. Therefore failure rates of engineered components were not required, as they are now, for use in prediction techniques and consequently there was little incentive for the formal collection of failure data.

Another factor is that, until well into this century, component parts were individually fabricated in a 'craft' environment. Mass production and the attendant need for component standardization did not apply and the concept of a valid repeatable component failure rate could not exist. The reliability of each product was, therefore, highly dependent on the craftsman/manufacturer and less determined by the 'combination' of part reliabilities.

Nevertheless, mass production of standard mechanical parts has been the case since early in this century. Under these circumstances defective items can be identified readily, by means of inspection and test, during the manufacturing process, and it is possible to control reliability by quality-control procedures.

The advent of the electronic age, accelerated by the Second World War, led to the need for more complex mass-produced component parts with a higher degree of variability in the parameters and dimensions involved. The experience of poor field reliability of military equipment throughout the 1940s and 1950s focused attention on the need for more formal methods of reliability engineering. This gave rise to the collection of failure information from both the field and from the interpretation of test data. Failure rate data banks were created in the mid-1960s as a result of work at such organizations as UKAEA (UK Atomic Energy Authority) and RRE (Royal Radar Establishment, UK) and RADC (Rome Air Development Corporation US).

The manipulation of the data was manual and involved the calculation of rates from the incident data, inventories of component types and the records of elapsed hours. This activity was stimulated by the appearance of reliability prediction modelling techniques which require component failure rates as inputs to the prediction equations.

The availability and low cost of desktop personal computing (PC) facilities, together with versatile and powerful software packages, has permitted the listing and manipulation of incident data for an order less expenditure of working hours. Fast automatic sorting of the data encourages the analysis of failures into failure modes. This is no small factor in contributing to more effective reliability assessment, since generic failure rates permit only parts count reliability predictions. In order to address specific system failures it is necessary to input component failure modes into the fault tree or failure mode analyses.

The labour-intensive feature of data collection is the requirement for field recording which remains a major obstacle to complete and accurate information. Motivation of staff to provide field reports with sufficient relevant detail is a current management problem. The spread of PC facilities to this area will assist in that interactive software can be used to stimulate the required information input at the same time as other maintenance-logging activities.

With the rapid growth of built-in test and diagnostic features in equipment a future trend may be the emergence of some limited automated fault reporting.

Failure data have been published since the 1960s and each major document is described in Chapter 4.

1.2 HAZARDOUS FAILURES

In the early 1970s the process industries became aware that, with larger plants involving higher inventories of hazardous material, the practice of learning by mistakes was no longer acceptable. Methods were developed for identifying hazards and for quantifying the consequences of failures. They were evolved largely to assist in the decision-making process when developing or modifying plant. External pressures to identify and quantify risk were to come later.

By the mid-1970s there was already concern over the lack of formal controls for regulating those activities which could lead to incidents having a major impact on the health and safety of the general public. The Flixborough incident, which resulted in 28 deaths in June 1974, focused public and media attention on this area of technology. Many further events such as that at Seveso in Italy in 1976 right through to the more recent Piper Alpha offshore and Clapham rail incidents have kept that interest alive and resulted in guidance and legislation which are addressed in Chapters 19 and 20.

The techniques for quantifying the predicted frequency of failures were previously applied mostly in the domain of availability, where the cost of equipment failure was the prime concern. The tendency in the last few years has been for these techniques also to be used in the field of hazard assessment.

1.3 RELIABILITY AND RISK PREDICTION

System modelling, by means of failure mode analysis and fault tree analysis methods, has been developed over the last 20 years and now involves numerous software tools which enable predictions to be refined throughout the design cycle. The criticality of the failure rates of specific component parts can be assessed and, by successive computer runs, adjustments to the design configuration and to the maintenance philosophy can be made early in the design cycle in order to optimize reliability and availability. The need for failure rate data to support these predictions has thus increased and Chapter 4 examines the range of data sources and addresses the problem of variability within and between them.

In recent years the subject of reliability prediction, based on the concept of validly repeatable component failure rates, has become controversial. First, the extremely wide variability of failure rates of allegedly identical components under supposedly identical environmental and operating conditions is now acknowledged. The apparent precision offered by reliability-prediction models is thus not compatible with the accuracy of the failure rate parameter. As a result, it can be concluded that simplified assessments of rates and the use of simple models suffice. In any case, more accurate predictions can be both misleading and a waste of money.

The main benefit of reliability prediction of complex systems lies not in the absolute figure predicted but in the ability to repeat the assessment for different repair times, different redundancy arrangements in the design configuration and different values of component failure rate. This has been made feasible by the emergence of PC tools such as fault tree analysis packages, which permit rapid reruns of the prediction. Thus, judgements can be made on the basis of relative predictions with more confidence than can be placed on the absolute values.

Second, the complexity of modern engineering products and systems ensures that system failure does not always follow simply from component part failure. Factors such as:

- Failure resulting from software elements
- Failure due to human factors or operating documentation
- Failure due to environmental factors
- Common mode failure whereby redundancy is defeated by factors common to the replicated units

can often dominate the system failure rate.

The need to assess the integrity of systems containing substantial elements of software increased significantly during the 1980s. The concept of validly repeatable 'elements', within the software, which can be mapped to some model of system reliability (i.e. failure rate), is even more controversial than the hardware reliability prediction processes discussed above. The extrapolation of software test failure rates into the field has not yet established itself as a reliable modelling technique. The search for software metrics which enable failure rate to be predicted from measurable features of the code or design is equally elusive.

Reliability prediction techniques, however, are mostly confined to the mapping of component failures to system failure and do not address these additional factors. Methodologies are currently evolving to model common mode failures, human factors failures and software failures, but there is no evidence that the models which emerge will enjoy any greater precision than the existing reliability predictions based on hardware component failures. In any case the very thought process of setting up a reliability model is far more valuable than the numerical outcome.

DESIGN

Design
reliability

Duplication
Derating
Component Selection
Design Qualification
Equipment Diversity

MANUFACTURE

Change Control
Quality Assurance
Production Testing
Training
Method Study
Process Instructions

FIELD

Achieved
reliability

Failure Feedback
Replacement Strategy
Preventive Maintenance
User Interaction

Figure 1.1

Figure 1.1 illustrates the problem of matching a reliability or risk prediction to the eventual field performance. In practice, prediction addresses the component-based 'design reliability', and it is necessary to take account of the additional factors when assessing the integrity of a system.

In fact, Figure 1.1 gives some perspective to the idea of reliability growth. The 'design reliability' is likely to be the figure suggested by a prediction exercise. However, there will be many sources of failure in addition to the simple random hardware failures predicted in this way. Thus the 'achieved reliability' of a new product or system is likely to be an order, or even more, less than the 'design reliability'. Reliability growth is the improvement that takes place as modifications are made as a result of field failure information. A well established item, perhaps with tens of thousands of field hours, might start to approach the 'design reliability'. Section 12.3 deals with methods of plotting and extrapolating reliability growth.

1.4 ACHIEVING RELIABILITY AND SAFETY-INTEGRITY

Reference is often made to the reliability of nineteenth-century engineering feats. Telford and Brunel left us the Menai and Clifton bridges whose fame is secured by their continued existence but little is remembered of the failures of that age. If we try to identify the characteristics of design or construction which have secured their longevity then three factors emerge:

1. *Complexity*: The fewer component parts and the fewer types of material involved then, in general, the greater is the likelihood of a reliable item. Modern equipment, so often condemned for its unreliability, is frequently composed of thousands of component parts all of which interact within various tolerances. These could be called intrinsic failures, since they arise from a combination of drift conditions rather than the failure of a specific component. They are more difficult to predict and are therefore less likely to be foreseen by the designer. Telford's and Brunel's structures are not complex and are composed of fewer types of material with relatively well-proven modules.

2. *Duplication/replication*: The use of additional, redundant, parts whereby a single failure does not cause the overall system to fail is a frequent method of achieving reliability. It is probably the major design feature which determines the order of reliability that can be obtained. Nevertheless, it adds capital cost, weight, maintenance and power consumption. Furthermore, reliability improvement from redundancy often affects one failure mode at the expense of another type of failure. This is emphasised, in the next chapter, by an example.

3. *Excess strength*: Deliberate design to withstand stresses higher than are anticipated will reduce failure rates. Small increases in strength for a given anticipated stress result in substantial improvements. This applies equally to mechanical and electrical items. Modern commercial pressures lead to the optimization of tolerance and stress margins which just meet the functional requirement. The probability of the tolerance-related failures mentioned above is thus further increased.

The last two of the above methods are costly and, as will be discussed in Chapter 3, the cost of reliability improvements needs to be paid for by a reduction in failure and operating costs. This argument is not quite so simple for hazardous failures but, nevertheless, there is never an endless budget for improvement and some consideration of cost is inevitable.

We can see therefore that reliability and safety are 'built-in' features of a construction, be it mechanical, electrical or structural. Maintainability also contributes to the availability of a system, since it is the combination of failure rate and repair/down time which determines unavailability. The design and operating features which influence down time are also taken into account in this book.

Achieving reliability, safety and maintainability results from activities in three main areas:

1. *Design*:
   Reduction in complexity
   Duplication to provide fault tolerance
   Derating of stress factors
   Qualification testing and design review
   Feedback of failure information to provide reliability growth
2. *Manufacture*:
   Control of materials, methods, changes
   Control of work methods and standards
3. *Field use*:
   Adequate operating and maintenance instructions
   Feedback of field failure information
   Replacement and spares strategies (e.g. early replacement of items with a known wearout characteristic)

It is much more difficult, and expensive, to add reliability/safety after the design stage. The quantified parameters, dealt with in Chapter 2, must be part of the design specification and can no more be added in retrospect than power consumption, weight, signal-to-noise ratio, etc.

1.5 THE RAMS-CYCLE

The life-cycle model shown in Figure 1.2 provides a visual link between RAMS activities and a typical design-cycle. The top portion shows the specification and feasibility stages of design leading to conceptual engineering and then to detailed design.

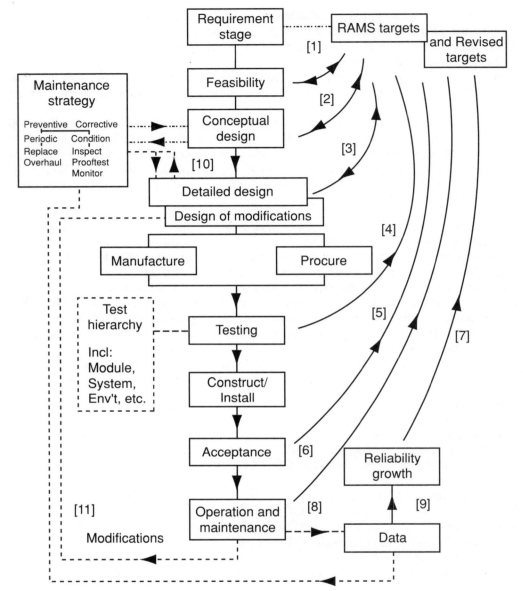

Figure 1.2 RAMS-Cycle model

RAMS targets should be included in the requirements specification as project or contractual requirements which can include both assessment of the design and demonstration of performance. This is particularly important since, unless called for contractually, RAMS targets may otherwise be perceived as adding to time and budget and there will be little other incentive, within the project, to specify them. Since each different system failure mode will be caused by different parts failures it is important to realize the need for separate targets for each undesired system failure mode.

Because one purpose of the feasibility stage is to decide if the proposed design is viable (given the current state-of-the-art) then the RAMS targets can sometimes be modified at that stage if initial predictions show them to be unrealistic. Subsequent versions of the requirements specification would then contain revised targets, for which revised RAMS predictions will be required.

The loops shown in Figure 1.2 represent RAMS related activities as follows:

- A review of the system RAMS feasibility calculations against the initial RAMS targets (loop [1]).
- A formal (documented) review of the conceptual design RAMS predictions against the RAMS targets (loop [2]).
- A formal (documented) review, of the detailed design, against the RAMS targets (loop [3]).
- A formal (documented) design review of the RAMS tests, at the end of design and development, against the requirements (loop [4]). This is the first opportunity (usually somewhat limited) for some level of real demonstration of the project/contractual requirements.
- A formal review of the acceptance demonstration which involves RAMS tests against the requirements (loop [5]). These are frequently carried out before delivery but would preferably be extended into, or even totally conducted, in the field (loop [6]).
- An ongoing review of field RAMS performance against the targets (loops [7,8,9]) including subsequent improvements.

Not every one of the above review loops will be applied to each contract and the extent of review will depend on the size and type of project.

Test, although shown as a single box in this simple RAMS-cycle model, will usually involve a test hierarchy consisting of component, module, subsystem and system tests. These must be described in the project documentation.

The maintenance strategy (i.e. maintenance programme) is relevant to RAMS since both preventive and corrective maintenance affect reliability and availability. Repair times influence unavailability as do preventive maintenance parameters. Loops [10] show that maintenance is considered at the design stage where it will impact on the RAMS predictions. At this point the RAMS predictions can begin to influence the planning of maintenance strategy (e.g. periodic replacements/overhauls, proof-test inspections, auto-test intervals, spares levels, number of repair crews).

For completeness, the RAMS-cycle model also shows the feedback of field data into a reliability growth programme and into the maintenance strategy (loops [8] [9] and [11]). Sometimes the growth programme is a contractual requirement and it may involve targets beyond those in the original design specification.

1.6 CONTRACTUAL PRESSURES

As a direct result of the reasons discussed above, it is now common for reliability parameters to be specified in invitations to tender and other contractual documents. Mean Times Between Failure, repair times and availabilities, for both cost- and safety-related failure modes, are specified and quantified.

There are problems in such contractual relationships arising from:

Ambiguity of definition
Hidden statistical risks
Inadequate coverage of the requirements
Unrealistic requirements
Unmeasurable requirements

Requirements are called for in two broad ways:

1. *Black box specification*: A failure rate might be stated and items accepted or rejected after some reliability demonstration test. This is suitable for stating a quantified reliability target for simple component items or equipment where the combination of quantity and failure rate makes the actual demonstration of failure rates realistic.
2. *Type approval*: In this case, design methods, reliability predictions during design, reviews and quality methods as well as test strategies are all subject to agreement and audit throughout the project. This is applicable to complex systems with long development cycles, and particularly relevant where the required reliability is of such a high order that even zero failures in a foreseeable time frame are insufficient to demonstrate that the requirement has been met. In other words, zero failures in ten equipment years proves nothing where the objective reliability is a mean time between failures of 100 years.

In practice, a combination of these approaches is used and the various pitfalls are covered in the following chapters of this book.

2 Understanding terms and jargon

2.1 DEFINING FAILURE AND FAILURE MODES

Before introducing the various Reliability parameters it is essential that the word *Failure* is fully defined and understood. Unless the failed state of an item is defined it is impossible to explain the meaning of Quality or of Reliability. There is only definition of failure and that is:

> Non-conformance to some defined performance criterion

Refinements which differentiate between terms such as Defect, Malfunction, Failure, Fault and Reject are sometimes important in contract clauses and in the classification and analysis of data but should not be allowed to cloud the issue. These various terms merely include and exclude failures by type, cause, degree or use. For any one specific definition of failure there is no ambiguity in the definition of reliability. Since failure is defined as departure from specification then revising the definition of failure implies a change to the performance specification. This is best explained by means of an example.

Consider Figure 2.1 which shows two valves in series in a process line. If the reliability of this 'system' is to be assessed, then one might enquire as to the failure rate of the individual valves. The response could be, say, 15 failures per million hours (slightly less than one failure per 7 years). One inference would be that the system reliability is 30 failures per million hours. However, life is not so simple.

If 'loss of supply' from this process line is being considered then the system failure rate is higher than for a single valve, owing to the series nature of the configuration. In fact it is double the failure rate of one valve. Since, however, 'loss of supply' is being specific about the requirement (or specification) a further question arises concerning the 15 failures per million hours. Do they all refer to the blocked condition, being the component failure mode which contributes to the system failure mode of interest? However, many failure modes are included in the 15 per million hours and it may well be that the failure rate for modes which cause 'no throughput' is, in fact, 7 per million hours.

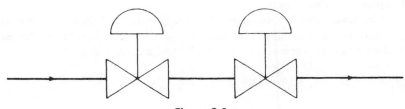

Figure 2.1

Table 2.1 Control valve failure rates per million hours

Fail shut	7
Fail open	3
Leak to atmosphere	2
Slow to move	2
Limit switch fails to operate	1
Total	15

Suppose, on the other hand, that one is considering loss of control leading to downstream over-pressure rather than 'loss of supply'. The situation changes significantly. First, the fact that there are two valves now enhances, rather than reduces, the reliability since, for this new system failure mode, both need to fail. Second, the valve failure mode of interest is the leak or fail open mode. This is another, but different, subset of the 15 per million hours – say, 3 per million. A different calculation is now needed for the system Reliability and this will be explained in Chapters 7 to 9. Table 2.1 shows a typical breakdown of the failure rates for various different failure modes of the control valve in the example.

The essential point in all this is that the definition of failure mode totally determines the system reliability and dictates the failure mode data required at the component level. The above example demonstrates this in a simple way, but in the analysis of complex mechanical and electrical equipment the effect of the defined requirement on the reliability is more subtle.

Given, then, that the word 'failure' is specifically defined, for a given application, quality and reliability and maintainability can now be defined as follows:

Quality: Conformance to specification.
Reliability: The probability that an item will perform a required function, under stated conditions, for a stated period of time. Reliability is therefore the extension of quality into the time domain and may be paraphrased as 'the probability of non-failure in a given period'.
Maintainability: The probability that a failed item will be restored to operational effectiveness within a given period of time when the repair action is performed in accordance with prescribed procedures. This, in turn, can be paraphrased as 'The probability of repair in a given time'.

2.2 FAILURE RATE AND MEAN TIME BETWEEN FAILURES

Requirements are seldom expressed by specifying values of reliability or of maintainability. There are useful related parameters such as Failure Rate, Mean Time Between Failures and Mean Time to Repair which more easily describe them. Figure 2.2 provides a model for the purpose of explaining failure rate.

The symbol for failure rate is λ (lambda). Consider a batch of N items and that, at any time t, a number k have failed. The cumulative time, T, will be Nt if it is assumed that each failure is replaced when it occurs whereas, in a non-replacement case, T is given by:

$$T = [t_1 + t_2 + t_3 \ldots t_k + (N - k)t]$$

where t_1 is the occurrence of the first failure, etc.

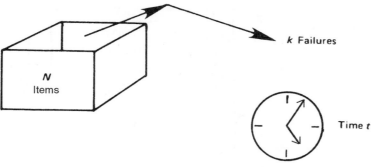

Figure 2.2

The Observed Failure Rate

This is defined: For a stated period in the life of an item, the ratio of the total number of failures to the total cumulative observed time. If λ is the failure rate of the N items then the observed λ is given by $\hat{\lambda} = k/T$. The $\wedge$ (hat) symbol is very important since it indicates that k/T is only an estimate of λ. The true value will be revealed only when all N items have failed. Making inferences about λ from values of k and T is the purpose of Chapters 5 and 6. It should also be noted that the value of $\hat{\lambda}$ is the average over the period in question. The same value could be observed from increasing, constant and decreasing failure rates. This is analogous to the case of a motor car whose speed between two points is calculated as the ratio of distance to time although the velocity may have varied during this interval. Failure rate is thus only meaningful for situations where it is constant.

Failure rate, which has the unit of t^{-1}, is sometimes expressed as a percentage per 1000 h and sometimes as a number multiplied by a negative power of ten. Examples, having the same value, are:

8500	per 10^9 hours (8500 FITS)
8.5	per 10^6 hours
0.85	per cent per 1000 hours
0.074	per year

Note that these examples each have only two significant figures. It is seldom justified to exceed this level of accuracy, particularly if failure rates are being used to carry out a reliability prediction (see Chapters 8 and 9).

The most commonly used base is per 10^6 h since, as can be seen in Appendices 3 and 4, it provides the most convenient range of coefficients from the 0.01 to 0.1 range for microelectronics, through the 1 to 5 range for instrumentation, to the tens and hundreds for larger pieces of equipment.

The per 10^9 base, referred to as FITS, is sometimes used for microelectronics where all the rates are small. The British Telecom database, HRD5, uses this base since it concentrates on microelectronics and offers somewhat optimistic values compared with other sources.

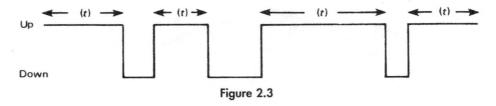

Figure 2.3

The Observed Mean Time Between Failures

This is defined: For a stated period in the life of an item the mean value of the length of time between consecutive failures, computed as the ratio of the total cumulative observed time to the total number of failures. If $\hat{\theta}$ (theta) is the MTBF of the N items then the observed MTBF is given by $\hat{\theta} = T/k$. Once again the hat indicates a point estimate and the foregoing remarks apply. The use of T/k and k/T to define $\hat{\theta}$ and $\hat{\lambda}$ leads to the inference that $\theta = 1/\lambda$.

This equality must be treated with caution since it is inappropriate to compute failure rate unless it is constant. It will be shown, in any case, that the equality is valid only under those circumstances. See Section 2.5, equations (2.5) and (2.6).

The Observed Mean Time to Fail

This is defined: For a stated period in the life of an item the ratio of cumulative time to the total number of failures. Again this is T/k. The only difference between MTBF and MTTF is in their usage. MTTF is applied to items that are not repaired, such as bearings and transistors, and MTBF to items which are repaired. It must be remembered that the time between failures excludes the down time. MTBF is therefore mean UP time between failures. In Figure 2.3 it is the average of the values of (t).

Mean life

This is defined as the mean of the times to failure where each item is allowed to fail. This is often confused with MTBF and MTTF. It is important to understand the difference. MTBF and MTTF can be calculated over any period as, for example, confined to the constant failure rate portion of the Bathtub Curve. Mean life, on the other hand, must include the failure of every item and therefore takes into account the wearout end of the curve. Only for constant failure rate situations are they the same.

To illustrate the difference between MTBF and life time compare:

- A match which has a short life but a high MTBF (few fail, thus a great deal of time is clocked up for a number of strikes)
- A plastic knife which has a long life (in terms of wearout) but a poor MTBF (they fail frequently)

2.3 INTERRELATIONSHIPS OF TERMS

Returning to the model in Figure 2.2, consider the probability of an item failing in the interval between t and $t + dt$. This can be described in two ways:

1. The probability of failure in the interval t to $t + dt$ given that it has survived until time t which is

   $$\lambda(t)\ dt$$

   where $\lambda(t)$ is the *failure rate*.

2. The probability of failure in the interval t to $t + dt$ unconditionally, which is

   $$f(t)\ dt$$

   where $f(t)$ is the *failure probability density function.*

The probability of survival to time t has already been defined as the reliability, $R(t)$. The rule of conditional probability therefore dictates that:

$$\lambda(t)\ dt = \frac{f(t)\ dt}{R(t)}$$

Therefore $$\lambda(t) = \frac{f(t)}{R(t)} \tag{2.1}$$

However, if $f(t)$ is the probability of failure in dt then:

$$\int_0^t f(t)\ dt = \text{probability of failure 0 to } t = 1 - R(t)$$

Differentiating both sides:

$$f(t) - = \frac{dR(t)}{dt} \tag{2.2}$$

Substituting equation (2.2) into equation (2.1),

$$-\lambda(t) = \frac{dR(t)}{dt} \cdot \frac{1}{R(t)}$$

Therefore integrating both sides:

$$-\int_0^t \lambda(t)\ dt = \int_1^{R(t)} dR(t)/R(t)$$

A word of explanation concerning the limits of integration is required. $\lambda(t)$ is integrated with respect to time from 0 to t. $1/R(t)$ is, however, being integrated with respect to $R(t)$. Now when $t = 0$, $R(t) = 1$ and at t the reliability $R(t)$ is, by definition, $R(t)$. Integrating then:

$$-\int_0^t \lambda(t)\ dt = \log_e R(t) \Big|_1^{R(t)}$$
$$= \log_e R(t) - \log_e 1$$
$$= \log_e R(t)$$

But if $a = e^b$ then $b = \log_e a$, so that:

$$R(t) = \exp\left[-\int_0^t \lambda(t)\, dt\right] \tag{2.3}$$

If failure rate is now assumed to be constant:

$$R(t) = \exp\left[-\int_0^t \lambda(t)\, dt\right] = \exp -\lambda t \Big|_0^t \tag{2.4}$$

Therefore $\underline{R(t) = e^{-\lambda t}}$

In order to find the MTBF consider Figure 2.3 again. Let $N - K$, the number surviving at t, be $N_s(t)$. Then $R(t) = N_s(t)/N$.

In each interval dt the time accumulated will be $N_s(t)\, dt$. At infinity the total will be

$$\int_0^\infty N_s(t)\, dt$$

Hence the MTBF will be given by:

$$\theta = \int_0^\infty \frac{N_s(t)\, dt}{N} = \int_0^\infty R(t)\, dt$$

$$\underline{\theta = \int_0^\infty R(t)\, dt} \tag{2.5}$$

This is the general expression for MTBF and always holds. In the special case of $R(t) = e^{-\lambda t}$ then

$$\theta = \int_0^\infty e^{-\lambda t}\, d_t$$

$$\underline{\theta = \frac{1}{\lambda}} \tag{2.6}$$

Note that inverting failure rate to obtain MTBF, and vice versa, is valid only for the constant failure rate case.

2.4 THE BATHTUB DISTRIBUTION

The much-used Bathtub Curve is an example of the practice of treating more than one failure type by a single classification. It seeks to describe the variation of Failure Rate of components during their life. Figure 2.4 shows this generalized relationship as originally assumed to apply to electronic components. The failures exhibited in the first part of the curve, where failure rate is decreasing, are called early failures or infant mortality failures. The middle portion is referred

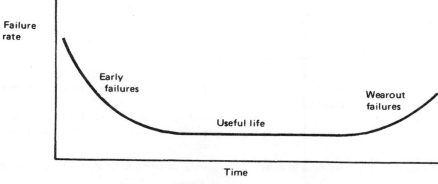

Figure 2.4

to as the useful life and it is assumed that failures exhibit a constant failure rate, that is to say they occur at random. The latter part of the curve describes the wearout failures and it is assumed that failure rate increases as the wearout mechanisms accelerate.

Figure 2.5, on the other hand, is somewhat more realistic in that it shows the Bathtub Curve to be the sum of three separate overlapping failure distributions. Labelling sections of the curve as wearout, burn-in and random can now be seen in a different light. The wearout region implies only that wearout failures predominate, namely that such a failure is more likely than the other types. The three distributions are described in Table 2.2.

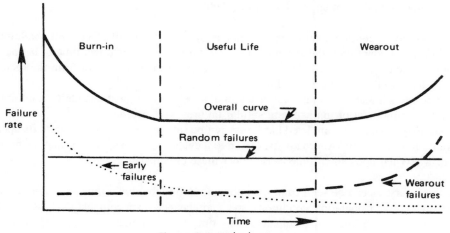

Figure 2.5 Bathtub Curve

2.5 DOWN TIME AND REPAIR TIME

It is now necessary to introduce Mean Down Time and Mean Time to Repair (MDT, MTTR). There is frequently confusion between the two and it is important to understand the difference. Down time, or outage, is the period during which equipment is in the failed state. A formal

Table 2.2

	Known as	
Decreasing failure rate	Infant mortality Burn-in Early failures	Usually related to manufacture and QA, e.g. welds, joints, connections, wraps, dirt, impurities, cracks, insulation or coating flaws, incorrect adjustment or positioning. In other words, populations of substandard items owing to microscopic flaws.
Constant failure rate	Random failures Useful life Stress-related failures Stochastic failures	Usually assumed to be stress-related failures. That is, random fluctuations (transients) of stress exceeding the component strength (see Chapter 11). The design reliability referred to in Figure 1.1 is of this type.
Increasing failure rate	Wearout failures	Owing to corrosion, oxidation, breakdown of insulation, atomic migration, friction wear, shrinkage, fatigue, etc.

definition is usually avoided, owing to the difficulties of generalizing about a parameter which may consist of different elements according to the system and its operating conditions. Consider the following examples which emphasize the problem:

1. A system not in continuous use may develop a fault while it is idle. The fault condition may not become evident until the system is required for operation. Is down time to be measured from the incidence of the fault, from the start of an alarm condition, or from the time when the system would have been required?
2. In some cases it may be economical or essential to leave equipment in a faulty condition until a particular moment or until several similar failures have accrued.
3. Repair may have been completed but it may not be safe to restore the system to its operating condition immediately. Alternatively, owing to a cyclic type of situation it may be necessary to delay. When does down time cease under these circumstances?

It is necessary, as can be seen from the above, to define the down time as required for each system under given operating conditions and maintenance arrangements. MTTR and MDT, although overlapping, are not identical. Down time may commence before repair as in (1) above. Repair often involves an element of checkout or alignment which may extend beyond the outage. The definition and use of these terms will depend on whether availability or the maintenance resources are being considered.

The significance of these terms is not always the same, depending upon whether a system, a replicated unit or a replaceable module is being considered.

Figure 2.6 shows the elements of down time and repair time:

a. *Realization Time*: This is the time which elapses before the fault condition becomes apparent. This element is pertinent to availability but does not constitute part of the repair time.

Figure 2.6 Elements of down time and repair time

b. *Access Time*: This involves the time, from realization that a fault exists, to make contact with displays and test points and so commence fault finding. This does not include travel but the removal of covers and shields and the connection of test equipment. This is determined largely by mechanical design.

c. *Diagnosis Time*: This is referred to as fault finding and includes adjustment of test equipment (e.g. setting up a lap top or a generator), carrying out checks (e.g. examining waveforms for comparison with a handbook), interpretation of information gained (this may be aided by algorithms), verifying the conclusions drawn and deciding upon the corrective action.

d. *Spare part procurement*: Part procurement can be from the 'tool box', by cannibalization or by taking a redundant identical assembly from some other part of the system. The time taken to move parts from a depot or store to the system is not included, being part of the logistic time.

e. *Replacement Time*: This involves removal of the faulty LRA (Least Replaceable Assembly) followed by connection and wiring, as appropriate, of a replacement. The LRA is the replaceable item beyond which fault diagnosis does not continue. Replacement time is largely dependent on the choice of LRA and on mechanical design features such as the choice of connectors.

f. *Checkout Time*: This involves verifying that the fault condition no longer exists and that the system is operational. It may be possible to restore the system to operation before completing the checkout in which case, although a repair activity, it does not all constitute down time.

g *Alignment Time*: As a result of inserting a new module into the system adjustments may be required. As in the case of checkout, some or all of the alignment may fall outside the down time.

h. *Logistic Time*: This is the time consumed waiting for spares, test gear, additional tools and manpower to be transported to the system.

i. *Administrative Time*: This is a function of the system user's organization. Typical activities involve failure reporting (where this affects down time), allocation of repair tasks, manpower changeover due to demarcation arrangements, official breaks, disputes, etc.

Activities (b)–(g) are called Active Repair Elements and (h) and (i) Passive Repair Activities. Realization time is not a repair activity but may be included in the MTTR where down time is the consideration. Checkout and alignment, although utilizing manpower, can fall outside the down time. The Active Repair Elements are determined by design, maintenance arrangements, environment, manpower, instructions, tools and test equipment. Logistic and Administrative time is mainly determined by the maintenance environment, that is, the location of spares, equipment and manpower and the procedure for allocating tasks.

Another parameter related to outage is Repair rate (μ). It is simply the down time expressed as a rate, therefore:

$$\mu = 1/\text{MTTR}$$

2.6 AVAILABILITY

In Chapter 1 Availability was introduced as a useful parameter which describes the amount of available time. It is determined by both the reliability and the maintainability of the item. Returning to Figure 2.3 it is the ratio of the (t) values to the total time. Availability is, therefore:

$$A = \frac{\text{Up time}}{\text{Total time}}$$

$$= \frac{\text{Up time}}{\text{Up time} + \text{Down time}}$$

$$= \frac{\text{Average of } (t)}{\text{Average of } (t) + \text{Mean down time}}$$

$$= \frac{\text{MTBF}}{\text{MTBF} + \text{MDT}}$$

This is know as the steady-state availability and can be expressed as a ratio or as a percentage. Sometimes it is more convenient to use Unavailability:

$$\overline{A} = 1 - A = \frac{\lambda \text{ MDT}}{1 + \lambda \text{ MDT}} \simeq \lambda \text{ MDT}$$

2.7 HAZARD AND RISK-RELATED TERMS

Failure rate and MTBF terms, such as have been dealt with in this chapter, are equally applicable to hazardous failures. *Hazard* is usually used to describe a situation with the potential for injury or fatality whereas *failure* is the actual event, be it hazardous or otherwise. The term *major hazard* is different only in degree and refers to certain large-scale potential incidents. These are dealt with in Chapters 10, 21 and 22.

Risk is a term which actually covers two parameters. The first is the probability (or rate) of a particular event. The second is the scale of consequence (perhaps expressed in terms of fatalities). This is dealt with in Chapter 10. Terms such as *societal* and *individual* risk differentiate between failures which cause either multiple or single fatalities.

2.8 CHOOSING THE APPROPRIATE PARAMETER

It is clear that there are several parameters available for describing the reliability and maintainability characteristics of an item. In any particular instance there is likely to be one parameter more appropriate than the others. Although there are no hard-and-fast rules the following guidelines may be of some assistance:

Failure Rate: Applicable to most component parts. Useful at the system level, whenever constant failure rate applies, because it is easy to compute Unavailability from $\lambda \times$ MDT. Remember, however, that failure rate is meaningless if it is not constant. The failure distribution would then be described by other means which will be explained in Chapter 6.
MTBF and MTTF: Often used to describe equipment or system reliability. Of use when calculating maintenance costs. Meaningful even if the failure rate is not constant.
Reliability/Unreliability: Used where the probability of failure is of interest as, for example, in aircraft landings where safety is the prime consideration.
Maintainability: Seldom used as such.
Mean Time To Repair: Often expressed in percentile terms such as the 95 percentile repair time shall be 1 hour. This means that only 5% of the repair actions shall exceed 1 hour.
Mean Down Time: Used where the outage affects system reliability or availability. Often expressed in percentile terms.
Availability/Unavailability: Very useful where the cost of lost revenue, owing to outage, is of interest. Combines reliability and maintainability. Ideal for describing process plant.
Mean Life: Beware of the confusion between MTTF and Mean Life. Whereas the Mean Life describes the average life of an item taking into account wearout, the MTTF is the average time between failures. The difference is clear if one considers the simple example of the match.

There are sources of standard definitions such as:

    BS 4778: Part 3.2
    BS 4200: Part 1
    IEC Publication 271
    US MIL STD 721B
    UK Defence Standard 00-5 (Part 1)
    Nomenclature for Hazard and Risk in the Process Industries (I Chem E)
    IEC 61508 (Part 4)

It is, however, not always desirable to use standard sources of definitions so as to avoid specifying the terms which are needed in a specification or contract. It is all too easy to 'define' the terms by calling up one of the aforementioned standards. It is far more important that terms are fully understood before they are used and if this is achieved by defining them for specific situations, then so much the better. The danger in specifying that all terms shall be defined by a given published standard is that each person assumes that he or she knows the meaning of each term and these are not read or discussed until a dispute arises. The most important area involving definition of terms is that of contractual involvement where mutual agreement as to the meaning of terms is essential. Chapter 19 will emphasize the dangers of ambiguity.

Useful notes

1. If failure rate is constant and, hence $R = e^{-\lambda t} = e^{-t/\theta}$, then after one MTBF the probability of survival, $R(t)$ is e^{-1}, which is 0.37.

2. If t is small, $e^{-\lambda t}$ approaches $1 - \lambda t$. For example, if $\lambda = 10^{-5}$ and $t = 10$ then $e^{-\lambda t}$ approaches $1 - 10^{-4} = 0.9999$.

3. Since $\theta = \int_0^\infty R(t)\, dt$, it is useful to remember that $\int_0^\infty A e^{-B\lambda t} = (A/B\lambda)$.

EXERCISES

If $\lambda = $ a) $1\ 10^{-6}$ per hr   b) $100\ 10^{-6}$ per hr

1. Calculate the MTBFs in years.
2. Calculate the Reliability for 1 year ($R_{(1yr)}$).
3. If the MDT is 10 hrs, calculate the Unavailability.
4. If the MTTR is 1 hour, the failures are dormant, and the inspection interval is 6 months, calculate the Unavailability.
5. What is the effect of doubling the MTTR?
6. What is the effect of doubling the inspection interval?

3 A cost-effective approach to quality, reliability and safety

3.1 THE COST OF QUALITY

The practice of identifying quality costs is not new, although it is only very large organizations that collect and analyse this highly significant proportion of their turnover. Attempts to set budget levels for the various elements of quality costs are even rarer. This is unfortunate, since the contribution of any activity to a business is measured ultimately in financial terms and the activities of quality, reliability and maintainability are no exception. If the costs of failure and repair were more fully reported and compared with the costs of improvement then greater strides would be made in this branch of engineering management. Greater recognition leads to the allocation of more resources. The pursuit of quality and reliability for their own sake is no justification for the investment of labour, plant and materials.

Quality Cost analysis entails extracting various items from the accounts and grouping them under three headings:

Prevention Costs – costs of preventing failures.
Appraisal Costs – costs related to measurement.
Failure Costs – costs incurred as a result of scrap, rework, failure, etc.

Each of these categories can be broken down into identifiable items and Table 3.1 shows a typical breakdown of quality costs for a six-month period in a manufacturing organization. The totals are expressed as a percentage of sales, this being the usual ratio. It is known by those who collect these costs that they are usually under-recorded and that the failure costs obtained can be as little as a quarter of the true value. The ratios shown in Table 3.1 are typical of a manufacturing and assembly operation involving light machining, assembly, wiring and functional test of electrical equipment. The items are as follows:

Prevention Costs
Design Review – Review of new designs prior to the release of drawings.
Quality and Reliability Training – Training of QA staff. Q and R Training of other staff.
Vendor Quality Planning – Evaluation of vendors' abilities to meet requirements.
Audits – Audits of systems, products, processes.
Installation Prevention Activities – Any of these activities applied to installations and the commissioning activity.
Product Qualification – Comprehensive testing of a product against all its specifications prior to the release of final drawings to production. Some argue that this is an appraisal cost. Since

Table 3.1 Quality costs: 1 January 1999 to 30 June 1999 (sales £2 million)

	£'000	% of Sales
Prevention Costs		
Design review	0.5	
Quality and reliability training	2	
Vendor quality planning	2.1	
Audits	2.4	
Installation prevention activities	3.8	
Product qualification	3.5	
Quality engineering	3.8	
	18.1	0.91
Appraisal Costs		
Test and inspection	45.3	
Maintenance and calibration	2	
Test equipment depreciation	10.1	
Line quality engineering	3.6	
Installation testing	5	
	66.0	3.3
Failure Costs		
Design changes	18	
Vendor rejects	1.5	
Rework	20	
Scrap and material renovation	6.3	
Warranty	10.3	
Commissioning failures	5	
Fault finding in test	26	
	87.1	4.36
Total quality cost	171.2	8.57

it is prior to the main manufacturing cycle the author prefers to include it in Prevention since it always attracts savings far in excess of the costs incurred.

Quality Engineering – Preparation of quality plans, workmanship standards, inspection procedures.

Appraisal Costs

Test and Inspection – All line inspection and test activities excluding rework and waiting time. If the inspectors or test engineers are direct employees then the costs should be suitably loaded. It will be necessary to obtain, from the cost accountant, a suitable overhead rate which allows for the fact that the QA overheads are already reported elsewhere in the quality cost report.

Maintenance and Calibration – The cost of labour and subcontract charges for the calibration, overhaul, upkeep and repair of test and inspection equipment.

Test Equipment Depreciation – Include all test and measuring instruments.

Line Quality Engineering – That portion of quality engineering which is related to answering test and inspection queries.

Installation Testing – Test during installation and commissioning.

Failure Costs

Design Changes – All costs associated with engineering changes due to defect feedback.

Vendor Rejects – Rework or disposal costs of defective purchased items where this is not recoverable from the vendor.

Rework – Loaded cost of rework in production and, if applicable, test.

Scrap and Material Renovation – Cost of scrap less any reclaim value. Cost of rework of any items not covered above.

Warranty – Warranty: labour and parts as applicable. Cost of inspection and investigations to be included.

Commissioning Failures – Rework and spares resulting from defects found and corrected during installation.

Fault Finding in Test – Where test personnel cary out diagnosis over and above simple module replacement then this should be separated out from test and included in this item. In the case of diagnosis being carried out by separate repair operators then that should be included.

A study of the above list shows that reliability and maintainability are directly related to these items.

UK industry turnover is in the order of £150 thousand million. The total quality cost for a business is likely to fall between 4% and 15%, the average being somewhere in the region of 8%. Failure costs are usually approximately 50% of the total – higher if insufficient is being spent on prevention. It is likely then that about £6 thousand million was wasted in defects and failures. A 10% improvement in failure costs would release into the economy approximately

£600 million

Prevention costs are likely to be approximately 1% of the total and therefore £1½ thousand million.

In order to introduce a quality cost system it is necessary to:

Convince top management – Initially a quality cost report similar to Table 3.1 should be prepared. The accounting system may not be arranged for the automatic collection and grouping of the items but this can be carried out on a one-off basis. The object of the exercise is to demonstrate the magnitude of quality costs and to show that prevention costs are small by comparison with the total.

Collect and Analyse Quality Costs – The data should be drawn from the existing accounting system and no major change should be made. In the case of change notes and scrapped items the effort required to analyse every one may be prohibitive. In this case the total may be estimated from a representative sample. It should be remembered, when analysing change notes, that some may involve a cost saving as well as an expenditure. It is the algebraic total which is required.

Quality Cost Improvements – The third stage is to set budget values for each of the quality cost headings. Cost-improvement targets are then set to bring the larger items down to an acceptable level. This entails making plans to eliminate the major causes of failure. Those remedies which are likely to realize the greatest reduction in failure cost for the smallest outlay should be chosen first.

Things to remember about Quality Costs are:

- They are not a target for individuals but for the company.
- They do not provide a comparison between departments because quality costs are rarely incurred where they are caused.
- They are not an absolute financial measure but provide a standard against which to make comparisons. Consistency in their presentation is the prime consideration.

3.2 RELIABILITY AND COST

So far, only manufacturers' quality costs have been discussed. The costs associated with acquiring, operating and maintaining equipment are equally relevant to a study such as ours. The total costs incurred over the period of ownership of equipment are often referred to as *Life Cycle Costs*. These can be separated into:

Acquisition Cost – Capital cost plus cost of installation, transport, etc.
Ownership Cost – Cost of preventive and corrective maintenance and of modifications.
Operating Cost – Cost of materials and energy.
Administration Cost – Cost of data acquisition and recording and of documentation.

They will be influenced by:

Reliability – Determines frequency of repair.
  Fixes spares requirements.
  Determines loss of revenue (together with maintainability).
Maintainability – Affects training, test equipment, down time, manpower.
Safety Factors – Affect operating efficiency and maintainability.

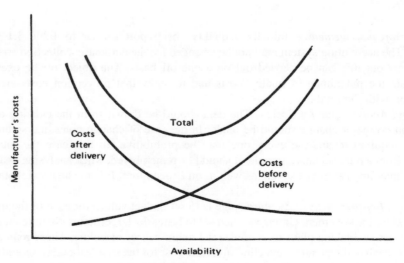

Figure 3.1 Availability and cost – manufacturer

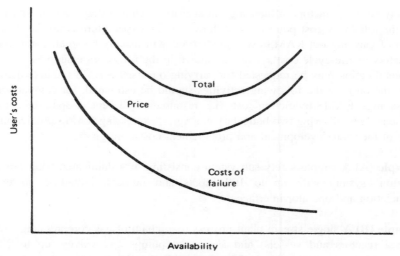

Figure 3.2 Availability and cost – user

Life cycle costs will clearly be reduced by enhanced reliability, maintainability and safety but will be increased by the activities required to achieve them. Once again the need to find an optimum set of parameters which minimizes the total cost is indicated. This concept is illustrated in Figures 3.1 and 3.2. Each curve represents cost against Availability. Figure 3.1 shows the general relationship between availability and cost. The manufacturer's pre-delivery costs, those of design, procurement and manufacture, increase with availability. On the other hand, the manufacturer's after-delivery costs, those of warranty, redesign, loss of reputation, decrease as availability improves. The total cost is shown by a curve indicating some value of availability at which minimum cost is incurred. Price will be related to this cost. Taking, then, the price/availability curve and plotting it again in Figure 3.2, the user's costs involve the addition of another curve representing losses and expense, owing to failure, borne by the user. The result is a curve also showing an optimum availability which incurs minimum cost. Such diagrams serve only to illustrate the philosophy whereby cost is minimized as a result of seeking reliability and maintainability enhancements whose savings exceed the initial expenditure.

A typical application of this principle is as follows:

- A duplicated process control system has a spurious shutdown failure rate of 1 per annum.
- Triplication reduces this failure rate to 0.8 per annum.
- The Mean Down Time, in the event of a spurious failure, is 24 hours.
- The total cost of design and procurement for the additional unit is £60 000.
- The cost of spares, preventive maintenance, weight and power arising from the additional unit is £1000 per annum.
- The continuous process throughput, governed by the control system, is £5 million per annum.
- The potential saving is $(1 - 0.8) \times 1/365 \times$ £5 million per annum = £2740 per annum which is equivalent to a capital investment of, say, <u>£30 000</u>
- The cost is £60 000 plus £1000 per annum which is equivalent to a capital investment of, say, <u>£70 000</u>

There may be many factors influencing the decision such as safety, weight, space available, etc. From the reliability cost point of view, however, the expenditure is not justified.

The cost of carrying out RAMS-cycle predictions will usually be small compared with the potential safety or life-cycle cost savings as shown in the following examples.

A cost justification may be requested for carrying out these RAMS prediction activities. In which case the costs of the following activities should be estimated, for comparison with the predicted savings. RAMS prediction costs (i.e. resources) will depend upon the complexity of the equipment. The following two budgetary examples, expressing RAMS prediction costs as a percentage of the total development and procurement costs, are given:

> **Example (A)** A simple safety subsystem consisting of a duplicated 'shut down' or 'fire detection' system with up to 100 inputs and outputs, including power supplies, annunciation and operator interfaces.

> **Example (B)** A single stream plant process (e.g. chain of gas compression, chain of H_2S removal reactors and vessels) and associated pumps and valves (up to 20) and the associated instrumentation (up to 50 pressure, flow and temperature transmitters).

	Man-days for (A)	Man-days for (B)
Figure 1.2 loop [1]: Feasibility RAMS prediction. This will consist of a simple block diagram prediction with the vessels or electronic controllers treated as units.	4	6
Figure 1.2 loop [2]: Conceptual design prediction. Similar to [1] but with more precise input/output quantities.	10	13
Figure 1.2 loop [3]: Detailed prediction. Includes FMECA at circuit level for 75% of the units, attention to common cause, human error and proof-test intervals.	6	18
Figure 1.2 loop [4]: RAMS testing. This refers to preparing subsystem and system test plans and analysis of test data rather then the actual test effort.	2	10
Figure 1.2 loop [5]: Acceptance testing. This refers to preparing test plans and analysis of test data rather then the actual test effort.	2	6
Figure 1.2 loop [6]: First year, reliability growth reviews. This is a form of design review using field data.	1	2
Figure 1.2 loop [7]: Subsequent reliability growth, data analysis.	2	3
Figure 1.2 loop [9]: First year, field data analysis. Not including effort for field data recording but analysis of field returns.	2	8
Figure 1.2 loop [10]: RCM planning. This includes identification of major components, establishing RAMS data for them, calculation of optimum discard, spares and proof-test intervals.	3	8

	Man-days for (A)	Man-days for (B)
Overall totals	32	74
Cost @ £250/man-day	£8K	£18.5K
Typical project cost (design and Procure)	£150K	£600K
RAMS cost as % of Total Project Cost	5.3%	3.1%

Life-cycle costs (for both safety and unavailability) can be orders greater than the above quoted project costs. Thus, even relatively small enhancements in MTBF/Availability will easily lead to costs far in excess of the example expenditures quoted above.

The cost of carrying out RAMS prediction activities is in the order of 3% to 5% of total project cost. Although definitive records are not readily available it is credible that the assessment process, with its associated comparison of alternatives and proposed modifications, will lead to savings which exceed this outlay. In the above examples, credible results of the RAM studies might be:

(A) ESD system:
The unavailability might typically be improved from 0.001 to 0.0005 as a result of the RAM study. Spurious shutdown, resulting from failure of the ESD, might typically be £500 000 per day for a small gas production platform. Thus, the **£8000** expenditure on RAM saves:

£500 000 × (0.001–0.0005) × 365 = **£91 000 per annum**

(B) H$_2$S system:
The availability might typically be improved from 0.95 to 0.98 as a result of the RAM study. Loss of throughput, resulting from failure, might typically cost £5000 per day. Thus, the **£18 500** expenditure on RAM saves:

£5000 × (0.98–0.95) × 365 = **£55 000 per annum**

Non RAMS-specialist engineers should receive training in RAMS techniques in order that they acquire sufficient competence to understand the benefits of those activities. The IEE/BCS competency guidelines document, 1999 offers a framework for assessing such competencies.

3.3 COSTS AND SAFETY

3.3.1 The need for optimization

Once the probability of a hazardous event has been assessed, the cost of the various measures which can be taken to reduce that risk is inevitably considered. If the risk to life is so high that it must be reduced as a matter of priority, or if the measures involved are a legal requirement, then the economics are of little or no concern – the equipment or plant must be made safe or closed down.

If, however, the risk to life is perceived to be sufficiently low then the reduction in risk for a given expenditure can be examined to see if the expenditure can be justified. At this point the concept of ALARP (As Low As Reasonably Practicable) becomes relevant in order that resources are allocated in the most effective manner. Risk has been defined as being ALARP if the reduction in risk is insignificant in relation to the cost of averting that risk. The problem here is that the words 'insignificant' and 'not worth the cost' are not quantifiably defined.

One approach is to consider the risks which society considers to be acceptable for both voluntary and involuntary situations. This is addressed in the Health and Safety Executive publications, *The Tolerability of Risk from Nuclear Power Installations* and *Reducing Risks Protecting People*, as well as some other publications in this area. This topic is developed in Section 10.2 of Chapter 10.

3.3.2 Cost per life saved

A controversial parameter is the Cost per Life Saved. This has a value in the ranking of possible expenditures so as to apply funds to the most effective area of risk improvement. Any technique which appears to put a price on human life is, however, potentially distasteful and thus attempts to use it are often resisted. It should not, in any case, be used as the sole criterion for deciding upon expenditure.

The concept is illustrated by the following hypothetical examples:

1. A potential improvement to a motor car braking system is costed at £40. Currently, the number of fatalities per annum in the UK is in the order of 3000. It is predicted that 500 lives per annum might be saved by the design. Given that 2 million cars are manufactured each year then the cost per life saved is calculated as:

$$\frac{£40 \times 2 \text{ million}}{500} = £160\,000$$

2. A major hazard process is thought to have an annual frequency of 10^{-6} for a release whose consequences are estimated to be 80 fatalities. An expenditure of £150 000 on new control equipment is predicted to improve this frequency to 0.8×10^{-6}. The cost per life saved, assuming a 40-year plant life, is thus:

$$\frac{£150\,000}{80 \times (10^{-6} - 0.8 \times 10^{-6}) \times 40} = £230 \text{ million}$$

The examples highlight the difference in acceptability between individual and societal risk. Many would prefer the expenditure on the major plant despite the fact that the vehicle proposal represents more lives saved per given expenditure. Naturally, such a comparison in cost per life saved terms would not be made, but the method has validity when comparing alternative approaches to similar situations.

The question arises as to the value of 'cost per life saved' to be used. Organizations are reluctant to state grossly disproportionate levels of CPL. Currently, figures in the range £500 000 to £2 000 000 are common. Where a risk has the potential for multiple fatalities then higher sums may be used.

However a value must be chosen, by the operator, for each assessment. The value selected must take account of any uncertainty inherent in the assessment and may have to take account of any company specific issues such as the number of similar installations. The greater the potential number of lives lost and the greater the aversion to the scenario then the larger is the choice of the cost per life saved criteria. Values which have been quoted include:

1. Approximately £1 000 000 by HSE, 1999 where there is a recognized scenario, a voluntary aspect to the exposure, a sense of having personal control, small numbers of casualties per incident. An example would be PASSENGER ROAD TRANSPORT.
2. Approximately £2 000 000–£4 000 000 by the HSE, 1991 where the risk is not under personal control and therefore an involuntary risk. An example would be TRANSPORT OF DANGEROUS GOODS.
3. Approximately £5 000 000–£15 000 000, mooted in the press, where there are large numbers of fatalities, there is uncertainty as to the frequency and no personal control by the victim. An example would be MULTIPLE RAIL PASSENGER FATALITIES.
4. This is a controversial area and figures can be subject to rapid revision in the light of catastrophic incidents and subsequent media publicity. A recent example, of the demand for automatic train protection in the UK, involves approximately £14 000 000 per life saved. This is despite the earlier rail industry practice of regarding £2 000 000 as an appropriate figure.

Part Two
Interpreting Failure Rates

4 Realistic failure rates and prediction confidence

4.1 DATA ACCURACY

There are many collections of failure rate data compiled by defence, telecommunications, process industries, oil and gas and other organizations. Some are published data Handbooks such as:

US MIL HANDBOOK 217 (Electronics)
CNET (French PTT) Data
HRD (Electronics, British Telecom)
RADC Non-Electronic Parts Handbook NPRD
OREDA (Offshore data)
FARADIP.THREE (Data ranges)

Some are data banks which are accessible by virtue of membership or fee such as:

SRD (Systems Reliability Department of UKAEA) Data Bank
Technis [the author] (Tonbridge)

Some are in-house data collections which are not generally available. These occur in:

Large industrial manufacturers
Public utilities

These data collection activities were at their peak in the 1980s but, sadly, they declined during the 1990s and the majority of published sources have not been updated since that time.

Failure data are usually, unless otherwise specified, taken to refer to random failures (i.e. constant failure rates). It is important to read, carefully, any covering notes since, for a given temperature and environment, a stated component, despite the same description, may exhibit a wide range of failure rates because:

1. Some failure rate data include items replaced during preventive maintenance whereas others do not. These items should, ideally, be excluded from the data but, in practice, it is not always possible to identify them. This can affect rates by an order of magnitude.
2. Failure rates are affected by the tolerance of a design and this will cause a variation in the values. Because definitions of failure vary, a given parametric drift may be included in one data base as a failure, but ignored in another.

3. Although nominal environmental and quality assurance levels are described in some databases, the range of parameters covered by these broad descriptions is large. They represent, therefore, another source of variability.
4. Component parts often are only described by reference to their broad type (e.g. signal transformer). Data are therefore combined for a range of similar devices rather than being separately grouped, thus widening the range of values. Furthermore, different failure modes are often mixed together in the data.
5. The degree of data screening will affect the relative numbers of intrinsic and induced failures in the quoted failure rate.
6. Reliability growth occurs where field experience is used to enhance reliability as a result of modifications. This will influence the failure rate data.
7. Trial and error replacement is sometimes used as a means of diagnosis and this can artificially inflate failure rate data.
8. Some data records undiagnosed incidents and 'no fault found' visits. If these are included in the statistics as faults, then failure rates can be inflated. Quoted failure rates are therefore influenced by the way they are interpreted by an analyst.

Failure rate values can span one or two orders of magnitude as a result of different combinations of these factors. Prediction calculations are explained in Chapters 8 and 9 and it will be seen (Section 4.4) that the relevance of failure rate data is more important than refinements in the statistics of the calculation. The data sources described in Section 4.2 can at least be subdivided into 'Site specific', 'Industry specific' and 'Generic' and the work described in Section 4.4 will show that the more specific the data source the greater the confidence in the prediction.

Failure rates are often tabulated, for a given component type, against ambient temperature and the ratio of applied to rated stress (power or voltage). Data is presented in one of two forms:

1. *Tables:* Lists of failure rates such as those in Appendices 3 and 4, with or without multiplying factors, for such parameters as quality and environment.
2. *Models:* Obtained by regression analysis of the data. These are presented in the form of equations which yield a failure rate as a result of inserting the device parameters into the appropriate expression.

Because of the large number of variables involved in describing microelectronic devices, data are often expressed in the form of models. These regression equations (WHICH GIVE A TOTALLY MISLEADING IMPRESSION OF PRECISION) involve some or all of the following:

- Complexity (number of gates, bits, equivalent number of transistors).
- Number of pins.
- Junction temperature (see Arrhenius, Section 11.2).
- Package (ceramic and plastic packages).
- Technology (CMOS, NMOS, bipolar, etc.).
- Type (memory, random LSI, analogue, etc.).
- Voltage or power loading.
- Quality level (affected by screening and burn-in).
- Environment.
- Length of time in manufacture.

Although empirical relationships have been established relating certain device failure rates to specific stresses, such as voltage and temperature, no precise formula exists which links specific environments to failure rates. The permutation of different values of environmental factors, such as those listed in Chapter 12, is immense. General adjustment (multiplying) factors have been evolved and these are often used to scale up basic failure rates to particular environmental conditions.

Because Failure Rate is, probably, the least precise engineering parameter, it is important to bear in mind the limitations of a Reliability prediction. The work described in Section 4.4 now makes it possible to express predictions using confidence intervals. The resulting MTBF, Availability (or whatever) should not be taken as an absolute parameter but rather as a general guide to the design reliability. Within the prediction, however, the relative percentages of contribution to the total failure rate are of a better accuracy and provide a valuable tool in design analysis.

Because of the differences between data sources, comparisons of reliability should always involve the same data source in each prediction.

For any reliability assessment to be meaningful it must address a specific system failure mode. To predict that a safety (shutdown) system will fail at a rate of, say, once per annum is, on its own, saying very little. It might be that 90% of the failures lead to a spurious shutdown and 10% to a failure to respond. If, on the other hand, the ratios were to be reversed then the picture would be quite different.

The failure rates, mean times between failures or availabilities must therefore be assessed for defined failure types (modes). In order to achieve this, the appropriate component level failure modes must be applied to the prediction models which are described in Chapters 8 and 9. Component failure mode data is sparse but a few of the sources do contain some information. The following sections indicate where this is the case.

4.2 SOURCES OF DATA

Sources of failure rate and failure mode data can be classified as:

1. SITE SPECIFIC
   Failure rate data which have been collected from similar equipment being used on very similar sites (e.g. two or more gas compression sites where environment, operating methods, maintenance strategy and equipment are largely the same). Another example would be the use of failure rate data from a flow corrector used throughout a specific distribution network. This data might be applied to the RAMS prediction for a new design of circuitry for the same application.
2. INDUSTRY SPECIFIC
   An example would be the use of the OREDA offshore failure rate data book for a RAMS prediction of a proposed offshore process package.
3. GENERIC
   A generic data source combines a large number of applications and sources.

As will be emphasized in Chapters 7–9, predictions require failure rates for specific modes of failure (e.g. open circuit, signal high, valve closes). Some, but unfortunately only a few, data sources contain specific failure mode percentages. Mean time to repair data is even more sparse although the OREDA data base is very informative in this respect.

The following are the more widely used sources:

4.2.1 Electronic failure rates

4.2.1.1 US Military Handbook 217 (Generic, no failure modes)

This is one of the better known data sources and is from RADC (Rome Air Data Centre in the USA). Opinions are sharply divided as to its value due to the unjustified precision implied by virtue of its regression model nature of its microelectronics sections. It covers:

Microelectronics
Discrete semiconductors
Tubes (thermionic)
Lasers
Resistors and capacitors
Inductors
Connections and connectors
Meters
Crystals
Lamps, fuses and other miscellaneous items

The Microelectronics sections present the information as a number of regression models. For example, the Monolithic Bipolar and MOS Linear Device model is given as:

Part operating failure rate model (λ_p):

$$\lambda p = \pi_Q (C_1 \pi_t\, \pi_V + C_2 \pi_E)\pi_L \text{ Failures}/10^6 \text{ hours}$$

where

π_Q is a multiplier for quality,
π_t is a multiplier for junction temperature,
π_V is a multiplier for applied voltage stress,
π_E is an application multiplier for environment,
π_L is a multiplier for the amount of time the device has been in production,
C_1 is based on the equivalent transistor count in the device,
C_2 is related to the packaging.

There are two reservations about this approach. First, it is not possible to establish the original application of the items from which the data are derived and it is not clear what mix of field and test data pertains. Second, a regression model both interpolates and extrapolates the results of raw data. There are similar models for other microelectronic devices and for discrete semiconductors. Passive components are described using tables of failure rates and the use of multipliers to take account of Quality and Environment.

The trend in successive issues of MIL 217 has been towards lower failure rates, particularly in the case of microelectronics. This is also seen in other data banks and may reflect the steady increase in manufacturing quality and screening techniques over the last 15 years. On the other hand, it may be due to re-assessing the earlier data. MIL 217 is available (as MILSTRESS) on disk from ITEM software. Between 1965 and 1991 it moved from Issue A to Issue F (amended 1992). It seems unlikely that it will be updated again.

4.2.1.2 HRD5 Handbook of Reliability Data for Electronic Components Used in Telecommunications Systems (Industry specific, no failure modes)

This document was produced, from field data, by British Telecom's Laboratories at Martlesham Heath and offers failure rate lists for Integrated Circuits, Discrete Semiconductors, Capacitors, Resistors, Electromechanical and Wound Components, Optoelectronics, Surge Protection, Switches, Visual Devices and a Miscellaneous section (e.g. microwave).

The failure rates obtained from this document are generally optimistic compared with the other sources, often by as much as an order of magnitude. This is due to an extensive 'screening' of the data whereby failures which can be attributed to a specific cause are eliminated from the data once remedial action has been introduced into the manufacturing process. Considerable effort is also directed towards eliminating maintenance-induced failures from the data.

Between 1977 and 1994 it moved from Issue 1 to Issue 5 but it seems unlikely that it will be updated again.

4.2.1.3 Recueil de Donnés de Fiabilité du CNET (Industry specific, no failure modes)

This document is produced by the Centre National d'Etudes des Telecommunications (CNET) now known as France Telecom R&D. It was first issued in 1981 and has been subject to subsequent revisions. It has a similar structure to US MIL 217 in that it consists of regression models for the prediction of component failure rates as well as generic tables. The models involve a simple regression equation with graphs and tables which enable each parameter to be specified. The model is also stated as a parametric equation in terms of voltage, temperature, etc. The French PTT use the CNET data as their standard.

4.2.1.4 BELLCORE, (Reliability Prediction Procedure for Electronic Equipment) TR-NWT–000332 Issue 5 1995 (Industry specific, no failure modes)

Bellcore is the research centre for the Bell telephone companies in the USA. Bellcore data is electronic failure rate data for telecommunications.

4.2.1.5 Electronic data NOT available for purchase

A number of companies maintain failure rate data banks including Nippon Telephone Corporation (Japan), Ericson (Sweden), and Thomson CSF (France) but this data is not generally available outside the organizations.

4.2.2 Other general data collections

Nonelectronic Parts Reliability data Book – NPRD (Generic, Some failure modes)

This document is also produced by RADC and was first published as: NPRD 1 in 1978 and was NPRD5 by 1995. It contains many hundreds of pages of failure rate information for a wide range of electromechanical, mechanical hydraulic and pneumatic parts. Failure rates are listed for a number of environmental applications. Unlike MIL 217, this is field data. It provides failure rate data against each component type and there are one or more entries per component type depending on the number of environmental applications for which a rate is available.

Each piece of data is given with the number of failures and hours (or operations/cycles). Thus there are frequently multiple entries for a given component type. Details for the breakdown of failure modes are given. NPRD 5 is available on disk.

4.2.2.2 OREDA – Offshore Reliability Data (1984/92/95/97) (Industry specific, Detailed failure modes, Mean times to repair)

This data book was prepared and published in 1984 and subsequently updated by a consortium of: BP Petroleum Development Ltd Norway, Elf Aquitaine Norge A/S, Norsk Agip A/S, A/S Norske Shell, Norsk Hydro a.s, Statoil, Saga Petroleum a.s and Total Oil Marine plc.

OREDA is managed by a steering committee made up from the participating companies. It is a collection of offshore failure rate and failure mode data with an emphasis on safety-related equipment. It covers components and equipment from:

    Fire and gas detection systems
    Process alarm systems
    Fire fighting systems
    Emergency shut down systems
    Pressure relieving systems
    General alarm and communication systems

OREDA 97 data is now available as a PC package, but only to members of the participating partners. The empty data base, however, is available for those wishing to collect their own data.

4.2.2.3 TECHNIS (the author) (Industry and generic, many failure modes, some repair times)

For 15 years, the author has collected a wide range of failure rate and mode data as well as recording the published data mentioned here. This is available to clients on a report basis. An examination of this data has revealed a 40% improvement between the 1980s and the 1990s.

4.2.2.4 UKAEA (Industry and generic, many failure modes)

This data bank is maintained by the Systems Reliability Department (SRD) of UKAEA at Warrington, Cheshire who have collected the data as a result of many years of consultancy. It is available on disk to members who pay an annual subscription.

4.2.2.5 Sources of nuclear generation data (Industry specific)

In the UK UKAEA, above, has some nuclear data, as has NNC (National Nuclear Corporation) although this may not be openly available.

In the USA Appendix III of the WASH 1400 study provided much of the data frequently referred to and includes failure rate ranges, event probabilities, human error rates and some common cause information. The IEEE standard IEEE500 also contains failure rates and restoration times. In addition there is NUCLARR (Nuclear Computerized Library for Assessing Reliability) which is a PC based package developed for the Nuclear Regulatory Commission and containing component failure rates and some human error data. Another US source is the NUREG publication. Some of the EPRI data is related to nuclear plant.

In France, Electricity de France provides the EIReDA mechanical and electrical failure rate data base which is available for sale.

In Sweden the TBook provides data on components in Nordic Nuclear Power Plants.

4.2.2.6 US sources of power generation data (Industry specific)

The EPRI (Electrical Power Research Institute) of GE Co., New York data scheme is largely gas turbine generation failure data in the USA.

There is also the GADS (Generating Availability Data System) operated by NERC (North American Electric Reliability Council). They produce annual statistical summaries based on experience from power stations in USA and Canada.

4.2.2.7 SINTEF (Industry specific)

SINTEF (at Trondheim) is part of the Norwegian Institute of Technology and, amongst many activities, collects failure rate data as, for example, data sheets on Fire and Gas Detection equipment.

4.2.2.8 Data not available for purchase

Many companies (e.g. Siemens), and for that matter firms of RAMS consultants (e.g. RM Consultants Ltd) maintain failure rate data but only for use by that organization.

4.2.3 Some older sources

A number of sources have been much used and are still frequently referred to. They are, however, somewhat dated but are listed here for completeness.

> Reliability Prediction Manual for Guided Weapon Systems (UK MOD) – DX99/013–100
> Reliability Prediction Manual for Military Avionics (UK MOD) – RSRE250
> UK Military Standard 00–41
> Electronic Reliability Data – INSPEC/NCSR (1981)
> Green and Bourne (book), *Reliability Technology*, Wiley 1972
> Frank Lees (book), *Loss Prevention in the Process Industries*, Butterworth-Heinemann.

4.3 DATA RANGES

For some components there is fairly close agreement between the sources and in other cases there is a wide range, the reasons for which were summarized in Section 4.1.

The FARADIP.THREE data base was created to show the ranges of failure rate for most component types. This database, CURRENTLY version 4.2 in 2001, is a summary of most of the other databases and shows, for each component, the range of failure rate values which is to be found from them. Where a value in the range tends to predominate then this is indicated. Failure mode percentages are also included. It is available on disk from the author at 26 Orchard Drive, Tonbridge, Kent TN10 4LG, UK and includes:

Discrete
> Diodes
> Opto-electronics
> Lamps and displays
> Crystals
> Tubes

Passive
> Capacitors
> Resistors
> Inductive
> Microwave

Instruments and Analysers
    Analysers
    Fire and Gas detection
    Meters
    Flow instruments
    Pressure instruments
    Level instruments
    Temperature instruments

Connection
    Connections and connectors
    Switches and breakers
    PCBs cables and leads

Electro-mechanical
    Relays and solenoids
    Rotating machinery (fans, motors, engines)

Power
    Cells and chargers
    Supplies and transformers

Mechanical
    Pumps
    Valves and parts
    Bearings
    Miscellaneous

Pneumatics
Hydraulics
Computers, data processing and communications
Alarms, fire protection, arresters and fuses

The ranges are presented in three ways:

1. *A single value*, where the various references are in good agreement.
2. *Two values indicating a range*. It is not uncommon for the range to be an order of magnitude wide. The user, as does the author, must apply engineering judgement in choosing a value. This involves consideration of the size, application and type of device in question. Where two values occupy the first and third columns then an even spread of failure rates is indicated. Where the middle and one other column are occupied then a spread with predominance to the value in the middle column is indicated.
3. *Three values indicating a range*. This implies that there is a fair amount of data available but that it spans more than an order of magnitude in range. Where the data tend to predominate in one area of the range then this is indicated in the middle column. The most likely explanation of the range widths is the fact that some data refer only to catastrophic failures whereas other data include degraded performance and minor defects revealed during preventive maintenance. This should be taken into account when choosing a failure rate from the tables.

As far as possible, the data given are for a normal ground fixed environment and for items procured to a good standard of quality assurance as might be anticipated from a reputable manufacturer operating to ISO 9000. The variation which might be expected due to other environments and quality arrangements is dealt with by means of multiplying factors.

SAMPLE FARADIP SCREEN – Fire and Gas Detection

	Failure rates, per million hours		
Gas pellister (fail 0.003)	5.00	10	30
Detector smoke ionization	1.00	6.00	40
Detector ultraviolet	5.00	8.00	20
Detector infra red (fail 0.003)	2.00	7.00	50
Detector rate of rise	1.00	4.00	12
Detector temperature	0.10	2.00	–
Firewire/rod + psu	25	–	–
Detector flame failure	1.00	10	200
Detector gas IR (fail 0.003)	1.50	5.00	80

Failure modes (proportion):

Rate of rise	Spurious 0.6	Fail 0.4
Temp, firewire/rod	Spurious 0.5	Fail 0.5
Gas pellister	Spurious 0.3	Fail 0.7
Infra red	Spurious 0.5	Fail 0.5
Smoke (ionize) and UV	Spurious 0.6	Fail 0.4

Using the ranges

The average range ratio for the entire FARADIP.THREE database is 7:1 In all cases, site specific failure rate data or even that acquired from identical (or similar) equipment, and being used under the same operating conditions and environment, should be used in place of any published data.

Such data should, nevertheless, be compared with the appropriate range. In the event that it falls outside the range there is a case for closer examination of the way in which the data were collected or in which the accumulated component hours were estimated.

Where the ranges contain a single value it can be used without need for judgement unless the specific circumstances of the assessment indicate a reason for a more optimistic or pessimistic failure rate estimate. Two or three values with predominating centre column: In the absence of any specific reason to favour the extreme values the predominating value is the most credible choice.

Where there are wide ranges with ratios >10:1 the use of the geometric mean is justified for the following reasons. The use of the simple arithmetic mean is not satisfactory for selecting a representative number when the two estimates are so widely spaced, since it favours the higher figure. The following example compares the arithmetic and geometric means where:

(1) Arithmetic Mean of n values of λ_i is given by

$$\sum_i^n \lambda_i/n$$

and (2) the Geometric Mean by:

$$\left(\prod_i^n \lambda_i\right)^{1/n}$$

Consider two estimates of failure rate, 0.1 and 1.0 (per million hours). The Arithmetic Mean (0.55) is five times the lower value and only a half of the upper value, thereby favouring the 1.0 failure rate. Where the range is an order or more, the larger value has significantly more bias on the arithmetic mean than the smaller.

The Geometric Mean (0.316) is, on the other hand, related to both values by a multiple of 3 and the excursion is thus the same. The Geometric Mean is, of course, derived from the Arithmetic Mean of the logarithms and therefore provides an average of the orders of magnitude involved. It is thus a more desirable parameter for describing the range.

In order to express the ranges as a single failure rate it is thus proposed to utilize the Geometric Mean. Appendix 3 shows microelectronic data in three columns giving the minima, maxima and geometric means. They can be interpreted as follows:

1. In general the lower figure in the range, used in a prediction, is likely to yield an assessment of the credible design objective reliability. That is the reliability which might reasonably be targeted after some field experience and a realistic reliability growth programme. The initial (field trial or prototype) reliability might well be an order of magnitude less than this figure.
2. The centre column figure indicates a failure rate which is more frequently indicated by the various sources. It is therefore a matter of judgement, depending on the type of prediction being carried out, as to whether it should be used in place of the lower figure.
3. The higher figure will probably include a high proportion of maintenance revealed defects and failures. The fact that data collection schemes vary in the degree of screening of maintenance revealed defects explains the wide ranges of quoted values.

4.4 CONFIDENCE LIMITS OF PREDICTION

The ratio of predicted failure rate (or system unavailability) to field failure rate (or system unavailability) was calculated for each of 44 examples and the results (part of the author's Ph.D. study) were classified in three categories:

(a) *Predictions using site specific data*: These are predictions based on failure rate data which have been collected from similar equipment being used on very similar sites (e.g. two or more sites where environment, operating methods, maintenance strategy and equipment are largely the same).
(b) *Predictions using industry specific data*: An example would be the use of the OREDA offshore failure rate data book for a RAMS prediction of a proposed offshore gas compression package.
(c) *Predictions using generic data*: These are predictions for which neither of the above two categories of data are available. Generic data sources (listed above) are used. FARADIP.THREE is also a generic data source in that it combines a large number of sources.

The results are:

1. For a prediction using site specific data

One can be this confident	That the eventual field failure rate will be BETTER than:
95%	$3\frac{1}{2}$ times the predicted
90%	$2\frac{1}{2}$ times the predicted
60%	$1\frac{1}{2}$ times the predicted
One can be this confident	That the eventual field failure rate will be in the range:
90%	$3\frac{1}{2}$:1 to 2/7:1

2. For a prediction using industry specific data

One can be this confident	That the eventual field failure rate will be BETTER than:
95%	5 times the predicted
90%	4 times the predicted
60%	$2\frac{1}{2}$ times the predicted
One can be this confident	That the eventual field failure rate will be in the range:
90%	5:1 to 1/5:1

3. For a prediction using generic data

One can be this confident	That the eventual field failure rate will be BETTER than:
95%	8 times the predicted
90%	6 times the predicted
60%	3 times the predicted
One can be this confident	That the eventual field failure rate will be in the range:
90%	8:1 to 1/8:1

Additional evidence in support of the 8:1 range is provided from the FARADIP data bank which suggests 7:1.

It often occurs that mixed data sources are used for a RAMS prediction such that, for example, site specific data are available for a few component parts but generic data are used for the other parts. The confidence range would then be assessed as follows:

If $Range_s$ and $Range_g$ are the confidence ranges for the site specific and generic data expressed as a multiplier then the range for a given prediction becomes

$$\frac{(\Sigma\lambda_s \times Range_s) + (\Sigma\lambda_g \times Range_g)}{\Sigma\lambda_s + \Sigma\lambda_g}$$

where $\Sigma\lambda_s$ and $\Sigma\lambda_g$ are the total failure rates of the site specific and generic items respectively.

For example, using the $3\frac{1}{2}$:1 and 8:1 ranges (90% confidence) given above, if $\Sigma\lambda_s = 20$ per million hrs (pmh) and $\Sigma\lambda_g = 100$ pmh, the range for the prediction (at 90% confidence) would be:

$$\frac{(20 \times 3.5) + (100 \times 8)}{120} = 7.25:1$$

At the end of Chapter 9 these ranges are used to compare predictions with targets.

4.5 OVERALL CONCLUSIONS

The use of stress-related regression models, implies an unjustified precision in estimating the failure rate parameter.

Site specific data should be used in preference to industry specific data which, in turn, should be used in preference to generic data.

Predictions should be expressed in confidence limit terms (Section 9.6) using the above information.

The FARADIP.THREE software package provides maximum and minimum rates together with failure modes.

In practice, failure rate is a system level effect. It is closely related to but not entirely explained by component failure. A significant proportion of failures encountered with modern electronic systems are not the direct result of parts failures but of more complex interactions within the system. The reason for this lack of precise mapping arises from such effects as human factors, software, environmental interference, interrelated component drift and circuit design tolerance.

The primary benefit to be derived from reliability engineering is the reliability growth which arises from continuing analysis and follow-up as well as corrective actions following failure analysis. Reliability prediction, based on the manipulation of failure rate data, involves so many potential parameters that a valid repeatable model for failure rate estimation is not possible. Thus, failure rate is the least accurate of engineering parameters and prediction from past data should be carried out either:

- As an indicator of the approximate level of reliability of which the design is capable, given reliability growth in the field
- To provide relative comparisons in order to make engineering decisions concerning optimum redundancy
- As a contractual requirement.
- In response to safety-integrity requirements

It should not be regarded as an accurate indicator of future field reliability.

5 Interpreting data and demonstrating reliability

5.1 THE FOUR CASES

In the following table it can be seen that there are four cases to be considered when interpreting the k failures and T hours. First, there may be reason to assume constant failure rate which includes two cases. If k is large (say, more than 10) then the sampling inaccuracy in such a wide-tolerance parameter may be ignored. Chapter 4 has emphasized the wide ranges which apply and thus, for large values of k the formulae:

$$\lambda = k/T \quad \text{and} \quad \theta = T/k$$

can be used. When k is small (even 0) the need arises to make some statistical interpretation of the data and that is the purpose of this chapter. The table also shows the second case where constant failure rate cannot be assumed. Again there may be few or many failures to interpret. Chapter 6 deals with this problem where the concept of a failure rate is not suitable to describe the failure distribution.

	CONSTANT FAILURE RATE	VARIABLE FAILURE RATE
FEW FAILURES	Chapter 5 (Statistical interpretation)	Chapter 6 (Inadequate data)
MANY FAILURES	Chapter 4 (Use $\lambda = k/T$)	Chapter 6 (Use probability plotting)

5.2 INFERENCE AND CONFIDENCE LEVELS

In Section 2.2 the concept of a point estimate of failure rate ($\hat{\lambda}$) or MTBF ($\hat{\theta}$) was introduced. In the model N items showed k failures in T cumulative hours and the observed MTBF ($\hat{\theta}$) of that sample measurement was T/k. If the test were repeated, and another value of T/k obtained, it would not be exactly the same as the first and, indeed, a number of tests would yield a number of values of MTBF. Since these estimates are the result of sampling they are called point

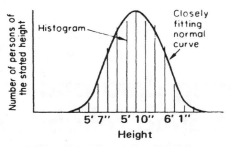

Figure 5.1 Distribution of heights

estimates and have the symbol $\hat{\theta}$. It is the true MTBF of the batch which is of interest and the only way to obtain it is to allow the entire batch to fail and then to evaluate T/k. This is why, the theoretical expression for MTBF in equation (2.5) of Section 2.3 involves the integration limits 0 and infinity:

$$\text{MTBF} = \int_0^\infty \frac{N_s(t)}{N}\, \mathrm{d}t$$

Thus, all devices must fail if the true MTBF is to be determined. Such a test will, of course, yield accurate data but, alas, no products at the end. In practice, we are forced to truncate tests after a given number of hours or failures. One is called a time-truncated test and the other a failure-truncated test. The problem is that a statement about MTBF, or failure rate, is required when only sample data are available. In many cases of high reliability the time required would be unrealistic.

The process of making a statement about a population of items based on the evidence of a sample is known as statistical inference. It involves, however, the additional concept of confidence level. This is best illustrated by means of an example. Figure 5.1 shows a distribution of heights of a group of people in histogram form. Superimposed onto the histogram is a curve of the normal distribution. The practice in statistical inference is to select a mathematical distribution which closely fits the data. Statements based on the distribution are then assumed to apply to the data.

In the figure there is a good fit between the normal curve, having a mean of 5'10" and a standard deviation (measure of spread) of 1", and the heights of the group in question. Consider, now, a person drawn, at random, from the group. It is permissible to state, from a knowledge of the normal distribution, that the person will be 5'10" tall or more providing that it is stated that the prediction is made with 50% confidence. This really means that we anticipate being correct 50% of the time if we continue to repeat the experiment. On this basis, an indefinite number of statements can be made, providing that an appropriate confidence level accompanies each value. For example:

5'11" or more at 15.9% confidence
6' 0" or more at   2.3% confidence
6' 1" or more at   0.1% confidence
OR between 5'9" and 5'11"   at   68.2% confidence

The inferred range of measurement and the confidence level can, hence, be traded off against each other.

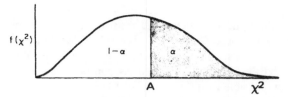

Figure 5.2 Single-sided confidence limits

5.3 THE CHI-SQUARE TEST

Returning to the estimates of MTBF, it is possible to employ the same technique of stating an MTBF together with a confidence level if the way in which the values are distributed is known. It has been shown that the expression

$$\frac{2k\hat{\theta}}{\theta} \text{ (random failures assumed)}$$

follows a χ^2 distribution with $2k$ degrees of freedom, where the test is truncated at the kth failure. We know already that

$$\hat{\theta} = \frac{T}{k} = \frac{\text{Accumulated test hours}}{\text{Number of failures}}$$

Therefore

$$\frac{2k\hat{\theta}}{\theta} = \frac{2kT}{k\theta} = \frac{2T}{\theta}$$

so that $2T/\theta$ is χ^2 distributed.

If a value of χ^2 can be fixed for a particular test then $2T/\theta$, and hence θ can be stated to lie between specified confidence limits. In practice, the upper limit is usually set at infinity and one speaks of an MTBF of some value or greater. This is known as the single-sided lower confidence limit of MTBF. Sometimes the double-sided limit method is used, although it is more usual to speak of the lower limit of MTBF.

It is not necessary to understand the following explanation of the χ^2 distribution. Readers who wish to apply the technique quickly and simply to the interpretation of data can move

DIRECTLY TO SECTION 5.5

For those who wish to understand the method in a little more detail then Figure 5.2 shows a distribution of χ^2. The area of the shaded portion is the probability of χ^2 exceeding that particular value at random.

In order to determine a value of χ^2 it is necessary to specify two parameters. The first is the number of degrees of freedom (twice the number of failures) and the second is the confidence level. The tables of χ^2 at the end of this book (Appendix 2) have columns and rows labelled α and n. The confidence level of the χ^2 distribution is α and n is the number of degrees of freedom. The limits of MTBF, however, are required between some value, A, and infinity. Since $\theta = 2T/\chi^2$

the value of χ^2 corresponding to infinite θ is zero. The limits are therefore zero and A. In Figure 5.2, if α is the area to the right of A then $1 - \alpha$ must be the confidence level of θ.

If the confidence limit is to be at 60%, the lower single-sided limit would be that value which the MTBF exceeds, by chance, six times out of 10. Since the degrees of freedom can be obtained from $2k$ and $\alpha = (1 - 0.6) = 0.4$, then a value of χ^2 can be obtained from the tables.

From $2T/\chi^2$ it is now possible to state a value of MTBF at 60% confidence. In other words, such a value of MTBF or better would be observed 60% of the time. It is written $\theta_{60\%}$. Alternatively $\lambda_{60\%} = \chi^2/2T$.

In a replacement test (each failed device is replaced immediately) 100 devices are tested for 1000 h during which three failures occur. The third failure happens at 1000 h at which point the test is truncated. We shall now calculate the MTBF of the batch at 90% and 60% confidence levels.

1. Since this is a replacement test T is obtained from the number under test multiplied by the linear test time. Therefore $T = 100\,000$ h and $k = 3$.
2. Let $n = 2k = 6$ degrees of freedom. For 90% confidence $\alpha = (1 - 0.9) = 0.1$ and for 60% confidence $\alpha = 1 - 0.6 = 0.4$.
3. Read off χ^2 values of 10.6 and 6.21 (see Appendix 2).
4. $\theta_{90\%} = 2 \times 100\,000/10.6 = 18\,900$ h.
   $\theta_{60\%} = 2 \times 100\,000/6.21 = 32\,200$ h.

Compare these results with the original point estimate of $T/k = 100\,000/3 = 33\,333$ h. It is possible to work backwards and discover what confidence level is actually applicable to this estimate. $\chi^2 = 2T/\theta = 200\,000/33\,333 = 6$. Since n is also equal to 6 it is possible to consult the tables and see that this occurs for a value of α slightly greater than 0.4. The confidence with which the MTBF may be quoted as $33\,333$ h is therefore less than 60%. It cannot be assumed that all point estimates will yield this value and, in any case, a proper calculation, as outlined, should be made.

In the above example the test was failure truncated. For a time-truncated test, one must be added to the number of failures (two to the degrees of freedom) for the lower limit of MTBF. This takes account of the possibility that, had the test continued for a few more seconds, a failure might have occurred. In the above single-sided test the upper limit is infinity and the value of MTBF is, hence, the lower limit. A test with zero failures can now be interpreted.

Consider 100 components for 50 h with no failures. At a 60% confidence we have $\theta_{60\%} = 2T/\chi^2 = 2 \times 50 \times 100/\chi^2$. Since we now have $\alpha = 0.4$ and $n = 2(k + 1) = 2$, $\chi^2 = 1.83$ and $\theta = 10\,000/1.83 = 5\,464$ h. Suppose that an MTBF of $20\,000$ h was required. The confidence with which it has been proved at this point is calculated as before. $\chi^2 = 2T/\theta = 20\,000/20\,000 = 1$. This occurs at $\alpha = 0.6$, therefore the confidence stands at 40%. If no failures occur then, as the test continues, the rise in confidence can be computed and observed. Furthermore, the length of the test (for zero failures) can be calculated in advance for a given MTBF and confidence level.

5.4 DOUBLE-SIDED CONFIDENCE LIMITS

So far, lower single-sided statements of MTBF have been made. Sometimes it is required to state that the MTBF lies between two confidence limits. Once again $\alpha = (1 - \text{confidence level})$ and is split equally on either side of the limits as shown in Figure 5.3.

Figure 5.3 Double-sided confidence limits

The two values of χ^2 are found by using the tables twice, first at $n = 2k$ and at $1 - \alpha/2$ (this gives the lower limit of χ^2) and second at $n = 2k$ ($2k + 2$ for time truncated) and at $\alpha/2$ (this gives the upper limit of χ^2). Once again, the upper limit of χ^2 corresponds with the lower limit of MTBF and vice versa. Figure 5.3 shows how $\alpha/2$ and $1 - \alpha/2$ are used. The probabilities of χ^2 exceeding the limits are the areas to the right of each limit and the tables are given accordingly.

Each of the two values of χ^2 can be used to obtain the limits of MTBF from the expression $\theta = 2T/\chi^2$. Assume that the upper and lower limits of MTBF for an 80% confidence band are required. In other words, limits of MTBF are required such that 80% of the time it will fall within them. $T = 100\ 000$ h and $k = 3$. The two values of χ^2 are obtained:

$n = 6, \alpha = 0.9, \chi^2 = 2.2$

$n = 8, \alpha = 0.1, \chi^2 = 13.4$

This yields the two values of MTBF – 14 925 h and 90 909 h, in the usual manner, from the expression $\theta = 2T/\chi^2$.

Hence the MTBF lies between 14 925 and 90 909 h with a confidence of 80%.

5.5 SUMMARIZING THE CHI-SQUARE TEST

The following list of steps summarizes the use of the χ^2 tables for interpreting the results of reliability tests:

1. Measure T (accumulated test hours) and k (number of failures).
2. Select a confidence level and let $\alpha = (1 - \text{confidence level})$.
3. Let $n = 2k$ ($2k + 2$ for lower limit MTBF in time-truncated test).
4. Note the value of χ^2 from the tables at the end of this book (Appendix 2).
5. Let MTBF at the given confidence level be $2T/\chi^2$ or $\lambda_{60\%} = \chi^2/2T$.
6. For double-sided limits use the above procedure twice at

$n = 2k : 1 - \alpha/2$ (upper limit of MTBF)
$n = 2k\ (2k + 2) : \alpha/2$ (lower limit of MTBF)

It should be noted that, for constant failure rate conditions, 100 components under test for 20 h yield the same number of accumulated test hours as 10 components for 200 h. Other methods

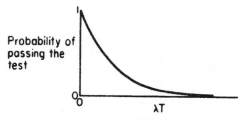

Figure 5.4 Zero failures test

of converting test data into statements of MTBF are available but the χ^2 distribution method is the most flexible and easy to apply. MTBFs are usually computed at the 60% and 90% confidence levels.

5.6 RELIABILITY DEMONSTRATION

Imagine that, as a manufactuer, you have evaluated the MTBF of your components at some confidence level using the techniques outlined, and that you have sold them to me on the basis of such a test. I may well return, after some time, and say that the number of failures experienced in a given number of hours yields a lower MTBF, at the same confidence, than did your earlier test. You could then suggest that I wait another month, by which time there is a chance that the number of failures and the number of test hours will have swung the calculation in your favour. Since this is hardly a suitable way of doing business it is necessary for consumer and producer to agree on a mutually acceptable test for accepting or rejecting batches of items. Once the test has been passed there is to be no question of later rejection on discovering that the batch passed on the strength of an optimistic sample. On the other hand, there is no redress if the batch is rejected, although otherwise acceptable, on the basis of a pessimistic sample. The risk that the batch, although within specification, will fail owing to a pessimistic sample being drawn is known as the producer's risk and has the symbol α (not to be confused with the α in the previous section). The risk that a 'bad' batch will be accepted owing to an optimistic sample is known as the consumer's risk, β. The test consists of accumulating a given number of test hours and then accepting or rejecting the batch on the basis of whether or not a certain number of failures have been observed.

Imagine such a test where the sample has to accumulate T test hours with no failures in order to pass. If the failure rate, λ, is assumed to be constant then the probability of observing no failures in T test hours is $e^{-\lambda T}$ (from the Poisson distribution). Such a zero failures test is represented in Figure 5.4, which is a graph of the probability of observing no failures (in other words, of passing the test) against the anticipated number of failures given by λT. This type of test is known as a Fixed Time Demonstration Test. It can be seen from the graph that, as the failure rate increases, the probability of passing the test falls.

The problem with this type of testing is the degree of *discrimination*. This depends on the statistical risks involved which are highlighted by the following example.

Assume, for the sake of argument, that the acceptable proportion of bad eggs (analogous to failure rate) is 10^{-4} (one in 10 000). If the reader were to purchase 6 eggs each week then he or she would be carrying out a demonstration test with a zero failures criterion. That is, with no bad eggs all is well, but if there is but one defective then a complaint will ensue. On the surface, this

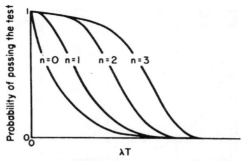

Figure 5.5 Family of OC curves

appears to be a valid test which carries a very high probability of being passed if the proportion of bad eggs is as stated.

Consider, however, the situation where the proportion increases to 10^{-3}, in other words by ten times. What of the test? The next purchase of 6 eggs is very unlikely to reveal a defect. This test is therefore a poor discriminator and the example displays, albeit lightheartedly, the problem of demonstrating a very high reliability (low failure rate). In many cases a statistical demonstration can be totally unrealistic for the reasons described above.

A component has an acceptable failure rate of 300×10^{-9}/h (approx. 1 in 380 yr). Fifty are tested for 1000 h (approx. 5½ years of test). λT is therefore

$$\frac{5.5}{380} = 0.014 \text{ and the probability of passing the test is } e^{-0.014} = 98.6\%.$$

Suppose that a second test is made from a batch whose failure rate is three times that of the first batch (i.e. 900×10^{-9}/h). Now the probability of passing the test is $e^{-\lambda T} = e^{-0.043} = 95.8\%$. Whereas the acceptable batch is 98.6% sure of acceptance ($\alpha = 1.4\%$) the 'bad' batch is only 4.2% sure of rejection ($\beta = 95.8\%$). In other words, although the test is satisfactory for passing batches of the required failure rate it is a poor discriminator whose acceptance probability does not fall sufficiently quickly as the failure rate increases.

A test is required which not only passes acceptable batches (a sensible producer's risk would be between 5% and 15%) but rejects batches with a significantly higher failure rate. Three times the failure rate should reduce the acceptance probability to 15% or less. The only way that this can be achieved is to increase the test time so that the acceptance criterion is much higher than zero failures (in other words, buy many more eggs!).

In general, the criterion for passing the test is n or fewer failures and the probability of passing the test is:

$$P_{0-n} = \sum_{i=0}^{n} \frac{\lambda^i T^i e^{-\lambda T}}{i!}$$

This expression yields the family of curves shown in Figure 5.5, which includes the special case ($n = 0$) of Figure 5.4. These curves are known as Operating Characteristics (OC Curves), each one representing a test plan.

Each of these curves represents a valid test plan and to demonstrate a given failure rate there is a choice of 0, 1, 2, 3, ..., n failure criterion tests with corresponding values of T. The higher

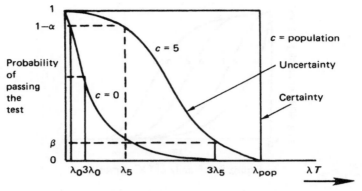

Figure 5.6 OC curves showing discrimination

the number of failures, the greater the number of test hours are required. Figure 5.6 shows the improvement in discrimination as n increases. Note that n is replaced by c, which is the usual convention. The extreme case where everything is allowed to fail and c equals the population is shown. Since there is no question of uncertainty under these circumstances, the probability of passing the test is either one or zero, depending upon the batch failure rate. The question of sampling risks does not arise.

Consider the $c = 0$ plan and note that a change from λ_0 to $3\lambda_0$ produces little decrease in the acceptance probability and hence a poor consumer's risk. If the consumer's risk were to be 10% the actual failure rate would be a long way to the right on the horizontal axis and would be many times λ_0. This ratio is known as the Reliability Design Index or Discrimination Ratio. Looking, now, at the $c = 5$ curve, both producer and consumer risks are reasonable for a 3:1 change in failure rate. In the extreme case of 100% failures both risks reduce to zero.

Figure 5.7 is a set of Cumulative Poisson Curves which enable the test plans and risks to be evaluated as in the following example.

A failure rate of 3×10^{-4}/h is to be demonstrated using 10 items. Calculate the number of test hours required if the test is to be passed with 4 or fewer failures and the probability of rejecting acceptable items (α) is to be 10%:

1. Probability of passing test $= 1 - 0.1 = 0.9$.
2. Using Figure 5.7 the corresponding value for $c = 4$ at 0.9 is 2.45.
3. $\lambda T = 3 \times 10^{-4} \times T = 2.45$. Therefore $T = 8170$ h.
4. Since there are 10 items the test must last 817 h with no more than four failures.

If the failure rate is three times the acceptable value calculate the consumer's risk, β:

1. $3\lambda T = 3 \times 3 \times 10^{-4} \times 8170 = 7.35$.
2. Using Figure 5.7 for $m = 7.35$ and $c = 4 : P_{0-4} = 0.15$.
3. The consumer's risk is therefore 15%.

Readers might care to repeat this example for a zero failures test and verify for themselves that, although T is as little as 333 h, β rises quickly to 74%. The difficulty of high-reliability testing can now be appreciated. For example, equipment which should have a one-year MTBF requires at least 3 years of testing to demonstrate its MTBF with acceptable risks. If only one item is available for test then the duration of the demonstration would be 3 years. In practice,

Figure 5.7 Poisson curves

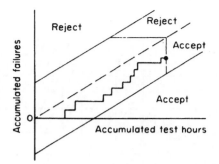

Figure 5.8 Truncated sequential demonstration test

far larger MTBFs are aimed for, particularly with submarine and satellite systems, and demonstration testing as described in this chapter is not applicable.

5.7 SEQUENTIAL TESTING

The above type of test is known as a Fixed-Time Demonstration. Owing to the difficulties of discrimination, any method that results in a saving of accumulated test hours without changing any of the other parameters is to be welcomed.

Experience shows that the Sequential Demonstration Test tends to achieve results slightly faster than the equivalent fixed-time test. Figure 5.8 shows how a sequential reliability test is operated. Two parallel lines are constructed so as to mark the boundaries of the three areas – Accept, Reject and Continue Testing. As test hours are accumulated the test proceeds along the *x*-axis and as failures occur the line is moved vertically one unit per failure. Should the test line cross the upper boundary, too many failures have been accrued for the hours accumulated and the test has been failed. If, on the other hand, the test crosses the lower boundary, sufficient test hours have been accumulated for the number of failures and the test has been passed. As long as the test line remains between the boundaries the test must continue.

Should a time limit be set to the testing then a truncating line is drawn as shown to the right of the diagram so that, if the line crosses above the mid-point, the test has been failed. If, as shown, it crosses below the mid-point, the test has been passed. If a decision is made by crossing the truncating line rather than one of the boundary lines, then the consumer and producer risks calculated for the test no longer apply and must be recalculated.

As in the fixed-time test, the consumer's risk, producer's risk and the MTBF associated with each are fixed. The ratio of the two MTBFs (or failure rates) is the reliability design index. The lines are constructed from the following equations:

$$y_{\text{upper}} = \frac{(1/\theta_1) - (1/\theta_0)}{\log_e(\theta_0/\theta_1)} T + \frac{\log_e A}{\log_e(\theta_0/\theta_1)}: \quad A \approx \frac{1 - \beta}{\alpha} \quad \text{and} \quad B \approx \frac{\beta}{1 - \alpha}$$

provided α and β are small (less than 25%).

The equation for y_{lower} is the same with $\log_e B$ substituted for $\log_e A$. If the risks are reduced then the lines move further apart and the test will take longer. If the design index is reduced, bringing the two MTBFs closer together, then the lines will be less steep, making it harder to pass the test.

5.8 SETTING UP DEMONSTRATION TESTS

In order to conduct a *demonstration test* (sometimes called a verification test) the following conditions, in addition to the statistical plans already discussed, must be specified:

1. Values of consumer's risk and acceptable MTBF. The manufacturer will then decide on the risk and upon a reliability design index. This has already been examined in this chapter. A failure distribution must be agreed (this chapter has dealt only with random failures). A test plan can then be specified.
2. The sampling procedure must be defined in terms of sample size and from where and how the samples should be drawn.
3. Both environmental and operational test conditions must be fixed. This includes specifying the location of the test and the test personnel.
4. Failure must be defined so that there will be no argument over what constitutes a failure once the test has commenced. Exceptions should also be defined, i.e. failures which are to be disregarded (failures due to faulty test equipment, wrong test procedures, etc.).
5. If a 'burn-in' period is to be allowed, in order that early failures may be disregarded, this too must be specified.

The emphasis in this chapter has been on component testing and demonstration, but if equipment or systems are to be demonstrated, the following conditions must also be specified:

1. Permissible corrective or preventive maintenance during the test (e.g. replacement of parts before wearout, routine care).
2. Relevance of secondary failures (failures due to fluctuations in stress caused by other failures).
3. How test time relates to real time (24 h operation of a system may only involve 3 h of operation of a particular unit).
4. Maximum setting-up and adjustment time permitted before the test commences.

US Military Standard 781C – Reliability Design Qualification and Production Acceptance Tests – contains both fixed-time and sequential test plans. Alternatively, plans can be easily constructed from the equations and curves given in this chapter.

EXERCISES

1. A replacement test involving 50 devices is run for 100 h and then truncated. Calculate the MTBF (single-sided lower limit) at 60% confidence:

   (a) If there are two failures;
   (b) If there are zero failures.

2. The items in Exercise 1 are required to show an MTBF of 5000 h at 90% confidence. What would be the duration of the test, with no failures, to demonstrate this?

3. The producer's risk in a particular demonstration test is set at 15%. How many hours must be accumulated, with no failures, to demonstrate an MTBF of 1000 h? What is the result if a batch is submitted to the test with an MTBF of 500 h? If the test were increased to five failures what would be the effect on T and β?

6 Variable failure rates and probability plotting

6.1 THE WEIBULL DISTRIBUTION

The Bathtub Curve in Figure 2.5 showed that, as well as random failures, there are distributions of increasing and decreasing failure rate. In these variable failure rate cases it is of little value to consider the actual failure rate since only Reliability and MTBF are meaningful. In Chapter 2 we saw that:

$$R(t) = \exp\left[-\int_0^t \lambda(t)\, dt\right]$$

Since the relationship between failure rate and time takes many forms, and depends on the device in question, the integral cannot be evaluated for the general case. Even if the variation of failure rate with time were known, it might well be of such a complicated nature that the integration would prove far from simple.

In practice it is found that the relationship can usually be described by the following three-parameter distribution known as the Weibull distribution named after Professor Waloddi Weibull:

$$R(t) = \exp\left[-\left(\frac{t-\gamma}{\eta}\right)^\beta\right]$$

In many cases a two parameter model proves sufficiently complex to describe the data. Hence:

$$R(t) = \exp\left[-(t/\eta)^\beta\right]$$

The constant failure rate case is the special one-parameter case of the Weibull distribution. Only randomness can be described by a single parameter.

In the general Weibull case the reliability function requires three parameters (γ, β, η) They do not have physical meanings in the same way as does failure rate. They are parameters which allow us to compute Reliability and MTBF. In the special case of $\gamma = 0$ and $\beta = 1$ the expression reduces to the exponential case with η giving the MTBF. In the general case, however, η is not the MTBF and is known as the scale parameter. β is known as the shape parameter and describes the rate of change of failure rate, increasing or decreasing. γ is known as the location parameter, in other words a displacement of the time origin. $\gamma = 0$ means that the time origin is, in fact, at $t = 0$.

The following equations show how data which can be described by a Weibull function can be made to fit a straight line. It is not essential to follow the explanation and the reader may, if desired, move to the next block of text.

The Weibull expression can be reduced to a straight-line equation by taking logarithms twice:

If $1 - R(t) = Q(t) \ldots$ the unreliability (probability of failure in t)

Then

$$1 - Q(t) = \exp\left[-\left(\frac{t - \gamma}{\eta}\right)^{\beta}\right]$$

so that

$$\frac{1}{1 - Q(t)} = \exp\left(\frac{t - \gamma}{\eta}\right)^{\beta}$$

Therefore

$$\log \frac{1}{1 - Q(t)} = \left(\frac{t - \gamma}{\eta}\right)^{\beta}$$

and

$$\log\log \frac{1}{1 - Q(t)} = \beta \log(t - \gamma) - \beta \log \eta$$

which is $Y = mX + C$, the equation of a straight line.

If $(t - \gamma)$ is replaced by t' then:

$$Y = \log\log \frac{1}{1 - Q(t)} \quad \text{and } X = \log t' \text{ and the slope } m = \beta.$$

If $Y = 0$

$$\log\log \frac{1}{1 - Q(t)} = 0$$

then

$$\beta \log t' = \beta \log \eta$$

so that

$$t' = \eta$$

This occurs if

$$\text{loglog} \; \frac{1}{1 - Q(t)} \; = \; 0 \quad \text{so that} \quad \text{log} \; \frac{1}{1 - Q(t)} \; = \; 1$$

i.e.

$$\frac{1}{1 - Q(t)} \; = \; e \quad \text{and} \quad Q(t) \; = \; 0.63$$

If a group of failures is distributed according to the Weibull function and it is initially assumed that $\gamma = 0$, then by plotting these failures against time on double logarithmic paper (failure percentage on loglog scale and time on log scale) a straight line should be obtained. The three Weibull parameters and hence the expression for Reliability may then be obtained from measurements of the slope and intercept.

Figure 6.1 is loglog by log graph paper with suitable scales for cumulative percentage failure and time. Cumulative percentage failure is effectively the unreliability and is estimated by taking each failure in turn from median ranking tables of the appropriate sample size. It should be noted that the sample size, in this case, is the number of failures observed. However, a test yielding 10 failures from 25 items would require the first 10 terms of the median ranking table for sample size 25.

6.2 USING THE WEIBULL METHOD

6.2.1 Curve fitting to interpret failure data

Assume that the failure rate is not constant **OR**, alternatively, that we want to determine whether it is or not.

Whereas, in the case of random failures (dealt with in Chapter 5) it was only necessary to know the total time T applying to the k failures, it is now necessary to know the individual times to failure of the items. Without this information it would not be possible to fit the data to a distribution.

The Weibull technique assumes, initially, that the distribution of failures, whilst not random, is at least able to be modelled by a simple 2 parameter distribution. It assumes that:

$$R(t) \; = \; \exp - (t/\eta)^\beta$$

The technique is to carry out a curve fitting (probability modelling) exercise to establish first that the data will fit this assumption and second to estimate the values of the 2 parameters.

Traditionally this has been done by 'pencil and paper' curve fitting methods which are described here. In a later section a software tool, for performing this task, is described.

If $\beta = 1$ then the failures are random and a constant failure rate can be assumed where failure rate $= 1/\eta$.

If $\beta > 1$ then the failure rate is increasing.

If $\beta < 1$ then the failure rate is decreasing.

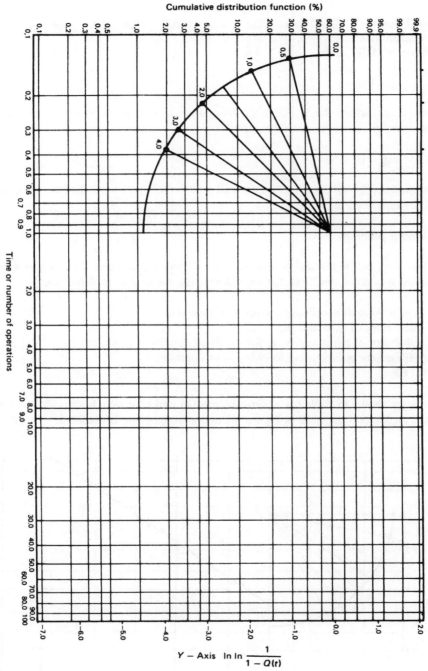

Figure 6.1 Graph paper for Weibull plot

Table 6.1

Cumulative failures, Q_t(%) median rank	6.7	16.2	25.9	35.6	45.2	54.8	64.5	74.1	83.8	93.3
Time, t (hours × 100)	1.7	3.5	5.0	6.4	8.0	9.6	11	13	18	22

In some cases, where the 2 parameter distribution is inadequate to model the data, the 3 parameter version can be used. In that case:

$$R(t) = \exp - [(t - \gamma)/\eta]^\beta$$

γ can be estimated by successive iteration until a fit to the 2 parameter distribution is obtained. This will be described in Section 6.3.

6.2.2 Manual plotting

Ten devices were put on test and permitted to fail without replacement. The time at which each device failed was noted and from the test information we require to determine:

1. If there is a Weibull distribution which fits these data;
2. If so, the values of γ, η and β;
3. The probability of items surviving for specified lengths of time;
4. If the failure rate is increasing, decreasing or constant;
5. The MTBF.

The results are shown in Table 6.1 against the median ranks for sample size 10. The ten points are plotted on Weibull paper as in Figure 6.2 and a straight line is obtained.

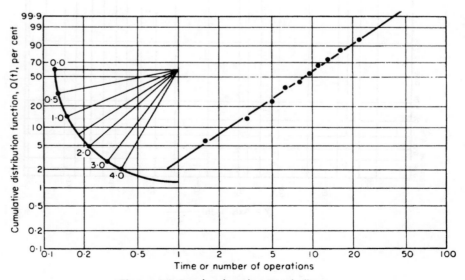

Figure 6.2 Results plotted on Weibull paper

The straight line tells us that the Weibull distribution is applicable and the parameters are determined as follows:

γ: It was shown in Section 6.1 that if the data yield a straight line then $\gamma = 0$.

β: The slope yields the value of β which is obtained by taking a line parallel to the data line but through the origin of the construction in Figure 6.2. The value of β is shown by the intersection with the arc. Here $\beta = 1.5$.

η: We have already shown that $\eta = t$ for $Q(t) = 0.63$, hence η is obtained by taking a horizontal line from the origin of the construction across to the data line and then reading the corresponding value of t.

The reliability expression is therefore:

$$R(t) = \exp\left[-\left(\frac{t}{1110}\right)^{1.5}\right]$$

The probability of survival to $t = 1000\,\text{h}$ is therefore:

$$R(1000) = e^{-0.855} = 42.5\%$$

The test shows a wearout situation since β, which is known as the shape parameter, >1.

For *increasing failure rate* $\beta > 1$
For *decreasing failure rate* $\beta < 1$
For *constant failure rate* $\beta = 1$

It now remains to evaluate the MTBF. This is, of course, the integral from zero to infinity of $R(t)$. Table 6.2 enables us to short-cut this step.

Since $\beta = 1.5$ then MTBF/$\eta = 0.903$ and MTBF $= 0.903 \times 1110 = 1002\,\text{h}$. Since median rank tables have been used the MTBF and reliability values calculated are at the 50% confidence

Table 6.2

β	$\dfrac{MTBF}{\eta}$	β	$\dfrac{MTBF}{\eta}$	β	$\dfrac{MTBF}{\eta}$	β	$\dfrac{MTBF}{\eta}$
0.0	∞	1.0	1.000	2.0	0.886	3.0	0.894
0.1	10!	1.1	0.965	2.1	0.886	3.1	0.894
0.2	5!	1.2	0.941	2.2	0.886	3.2	0.896
0.3	9.261	1.3	0.923	2.3	0.886	3.3	0.897
0.4	3.323	1.4	0.911	2.4	0.886	3.4	0.898
0.5	2.000	1.5	0.903	2.5	0.887	3.5	0.900
0.6	1.505	1.6	0.897	2.6	0.888	3.6	0.901
0.7	1.266	1.7	0.892	2.7	0.889	3.7	0.902
0.8	1.133	1.8	0.889	2.8	0.890	3.8	0.904
0.9	1.052	1.9	0.887	2.9	0.892	3.9	0.905
						4.0	0.906

level. In the example, time was recorded in hours but there is no reason why a more appropriate scale should not be used such as number of operations or cycles. The MTBF would then be quoted as Mean Number of Cycles between Failures.

For samples of other than 10 items a set of median ranking tables is required. Since space does not permit a full set to be included the following approximation is given. For sample size *N* the *r*th rank is obtained from Bernard's approximation:

$$\frac{r - 0.3}{N + 0.4}$$

Care must be taken in the choice of the appropriate ranking table. *N* is the number of items in the test and *r* the number that failed, in other words, the number of data points. In our example *N* was 10 not because the number of failures was 10 but because it was the sample size. As it happens, we considered the case where all 10 failed.

Had there been 20 items, of which 10 did not fail, the median ranks from Bernard's formula would have been:

%:– 3.4 8.3 13 18 23 28 33 38 43 48

Although this method allows for the ranking of the failures it does not take account of the actual hours contributed by the censored items. In the next section, the Maximum Likelihood technique is introduced partly for this purpose.

6.2.3 Using a computer method

The COMPARE software package provides a method of probability plotting whereby Weibull parameters are found which best fit the data being analysed.

Repair times and censored data are entered and estimates of the Weibull parameters, as well as a graphical plot, are provided.

There are four types of censoring:

– Items removed (for some reason other than failure) before the test finishes.
– Items which continue after the last failure.
– Items which are added after the commencement of the test whose operating hours count from their inclusion.
– Failed items which are restored to 'as new' condition and then clock up further operating time. Strictly speaking this is not an example of censoring since the item has been allowed to fail.

In the latter case it is important to be satisfied that the refurbishment really is 'as new'. If so the additional hours count from the refurbishment and are treated as an extra item.

In practice it may happen that there is a time to failure for a particular failure mode. The item might be repaired 'as new' and continue until it fails again. IMPORTANT – If the second failure is the same mode then the time to failure is counted from the refurbishment. If the second failure is a different mode then the time to failure is the whole operating time from the commencement of the test.

It MUST be remembered, however, that any computerized algorithm will allocate parameters to any data for a given distribution. It is, therefore, important to be aware of the limitations of probability plotting.

Two methods of estimating the Weibull parameters from a set of times to failure are LEAST SQUARES AND MAXIMUM LIKELIHOOD.

The Least Squares method is used as an initial calculation and involves calculating the hypothetical line for which the sum of the squares of the distances of the horizontal distances from the data points to the line is a minimum. The Weibull parameters, BETA and ETA, are obtained from the line. For the two parameter Weibull distribution the Least Squares estimates are obtained from:

$$\text{BETA} = (\Sigma(Y_i)^2 - \overline{Y} \, \Sigma Y_i)/(\Sigma X_i Y_i - \overline{X} \, \Sigma Y_i)$$

$$\text{ETA} = \text{Exp} \, (\overline{X} - \overline{Y}/\text{Beta})$$

where $Y = \text{Log}_e \, \{\text{log}_e[1/(1 - F(t))]\}$
$X = \text{Log}_e \, t$
$t = \text{time}$

Because this Least Squares method involves treating each of the squared distances with equal importance it favours the higher values of time. Nevertheless, the Least Squares estimates of BETA and ETA may well be adequate if there is very little, or better still, no censored data. However data sets usually involve some times to failure (the failed items) and some times with no failure (the survivors). In this case the MAXIMUM LIKELIHOOD estimate is required.

In COMPARE the Least Squares estimates of BETA and ETA are used as the most reasonable estimate for commencing the iterative process of determining Maximum Likelihood values which give equal weight to each data point by virtue of calculating its probability of causing the estimated parameter. The algorithm generates the Weibull BETA and ETA parameters from which the data are most likely to have come by setting up a likelihood equation, differentiating with respect to BETA and ETA, and setting this equal to zero (in other words the standard calculus method of obtaining a minimum). The process is iterated for alternate BETA and ETA estimates until the values do not significantly change.

The Maximum Likelihood values are then taken as the best estimates of the Weibull parameters.

A large number of data collection schemes do not readily provide the times to failure of the items in question. For example, if an assembly (such as a valve) is replaced from time to time then its identity and its time to failure and replacement might be obtainable from the data. However, it might well be the diaphragm which is eventually the item of interest. Diaphragms may have been replaced during routine maintenance and the identity of each diaphragm not recorded. Subsequent Weibull analysis of the valve diaphragm would not then be possible. Careful thought has to be given when implementing a data collection scheme as to what subsequent DATA ANALYSIS will take place.

As in the above example of a valve and its diaphragm each of SEVERAL FAILURE MODES will have its own failure distribution for which Weibull analysis may be appropriate. It is very likely, when attempting this type of modelling, that data not fitting the 2 parameter distribution actually contains more than one failure mode. Separating out the individual failure modes may permit successful Weibull modelling.

6.2.4 Significance of the result

The dangers of attempting to construct a Weibull plot with too few points should be noted. A satisfactory result will not be obtained with less than at least six points. Tests yielding 0, 1, 2 and even 3 failures do not allow a variable failure rate to be observed. In these cases constant failure rate must be assumed and the chi square test used which is a valid approach provided that the information extracted is applied only to the same time range as the test.

The comparison between the results obtained from Least Squares and Maximum Likelihood estimations (described above) provide an initial feel for how good a fit the data is to the inferred Weibull parameters.

If (in addition to the confidence obtained from the physical plot) the two values of Shape Parameter, obtained from Least Squares and Maximum Likelihood, are in good agreement there is a further test.

This is provided by way of the Gnedenko test which tests for constant failure rate. This is an 'F' test which tests the hypothesis that the failure times are at random, i.e. $\beta = 1$. The screen will state whether or not it is valid to reject the assumption that $\beta = 1$. The lower the value of the significance % then the more likely it is that the failure rate is significantly different from constant.

Essentially the test compares the MTTF of the failure times as grouped either side of the middle failure time and tests for a significant difference.

If the total number of failure times is n, and the time of the $n/2$th failure is T, the two estimates are:

$$\frac{\sum\limits_{i=1}^{n/2} t_i + \left(n/2 \times T \right)}{n/2} \quad \text{and} \quad \frac{\sum\limits_{i=n/2+1}^{n} (t_i - T)}{n/2}$$

That is to say we are comparing the MTTF of the 'first half' of the failures and the MTTF of the 'second half'. The ratio should be one if the failure rate is constant. If it is not then the magnitude of the ratio gives an indication of significance. The ratio follows an 'F' distribution and the significance level can therefore be calculated. The two values of MTTF are shown on the screen. If this test were applied to the graphical plot in Section 6.2.2, we would see that, despite a fairly good straight line, the confidence that β is not 1 is only 32%!

It should be remembered that a small number of failure times, despite a high value of β, may not show a significant departure from the 'random' assumption. In practice 10 or more failure times is a minimum desirable data set for Weibull analysis. Nevertheless, engineering judgement should always be used to temper statistical analysis. The latter looks only at numbers and does not take account of known component behaviours.

> *Note*: If a poor fit is obtained from the 2 parameter model, and the plot is a simple curve rather than 'S' shaped or disjointed, then it is possible to attempt a 3 parameter model by estimating the value of γ described in section 6.3. The usual approach is to assume that γ takes the value of the first failure time and to proceed, as above, with the 2 parameter model to find η and β. Successive values of γ can be attempted, by iteration, until the 2 parameter model provides a better fit. It must be remembered however that if the reason for a poor fit with the 2 parameter model is that only a few failure times are available then the use of the 3 parameter model is unlikely to improve the situation.

If the plot is 'S' shaped, then it is possible that two failure modes are present in the data.

In the author's experience only a limited number of components show a significantly increasing failure rate. This is often due to the phenomenon (known as Drenick's law) whereby a mixture of three or more failure modes will show a random failure distribution irrespective of the BETAs of the individual modes.

Table 6.3

Time, t (hours $\times$ 100)	1.7	3.5	5.0	6.4	8.0	9.6	11	13	18	22
Median rank	6.7	16.2	25.9	35.6	45.2	54.8	64.5	74.1	83.8	93.3
5% rank	0.5	3.7	8.7	15	22	30	39	49	61	74
95% rank	26	39	51	61	70	78	85	91	96	99

6.3 MORE COMPLEX CASES OF THE WEIBULL DISTRIBUTION

Suppose that the data in our example had yielded a curve rather than a straight line. It is still possible that the Weibull distribution applies but with γ greater than zero. The approach is to select an assumed value for γ, usually the first value of t in the data, and replot the line against t', where $t' = t - \gamma$. The first point is now not available and the line will be constructed from one point fewer. Should the result be a straight line then the value of γ is as estimated and one proceeds as before to evaluate the other two parameters. MTBF is calculated as before plus the value of γ. If, on the other hand, another curve is generated then a further value of γ is tried until, by successive approximations, the correct value is found. This trial and error method of finding γ is not as time consuming as it might seem. It is seldom necessary to attempt more than four approximations of γ before either generating a straight line or confirming that the Weibull distribution will not fit the data. One possible reason for the Weibull distribution not applying could be the presence of more than one failure mechanism in the data. Two mechanisms are unlikely to follow the same distribution and it is important to confine the analysis to one mechanism at a time.

So far, a single-sided analysis at 50% confidence has been described. It is possible to plot the 90% confidence bands by use of the 5% and 95% rank tables. First Table 6.3 is constructed and the confidence bands plotted as follows.

Figure 6.3 Ninety per cent confidence bands

Consider the point corresponding to the failure at 500 h. The two points A and B are marked on the straight line corresponding to 8.7% and 51% respectively. The median rank for this point was 25.9% and vertical lines are drawn from A and B to intersect the horizontal. These two points lie on the confidence bands. The other points are plotted in the same way and confidence bands are produced as shown in Figure 6.3. Looking at the curves the limits of $Q(t)$ at 1000 h are 30% and 85%. At 90% confidence the Reliability for 1000 h is therefore between 15% and 70%.

6.4 CONTINUOUS PROCESSES

There is a very strict limitation to the use of this Weibull method, which is illustrated by the case of filament lamps. It is well known that these do not fail at random. Indeed, they have a pronounced wearout characteristic with a β in excess of 2. However, imagine a brand-new building with brand-new lamps. Due to the distribution of failures, very few will fail in the first few months, perhaps only a few in the next few months and several towards the end of the year. After several years, however, the lamps in the building will all have been replaced at different times and the number failing in any month will be approximately the same. Thus, a population of items with increasing failure rate appears as a constant failure rate system. This is an example of a continuous process, and Figure 6.4 shows the failure characteristic of a single lamp and the superimposition of successive generations.

If the intervals between failure were observed, ranked and plotted in a Weibull analysis then a β of 1 would be obtained. Weibull analysis must not therefore be used for the times between failure within a continuous process but only for a number of items whose individual times to failure are separately recorded. It is not uncommon for people to attempt the former and obtain a totally false picture of the process.

One method of tackling this problem is to use the reliability growth models (CUSUM and Duane) described in Chapter 12. Another is to apply the Laplace Test which provides a means of indicating if the process failure rate has a trend.

If a system exhibits a number of failures after time zero at times $x_1,x_2,x_3,...,x_i$, then the test statistic for the process is

$$U = \frac{(\Sigma x_i/n) - (x_0/2)}{x_0 \sqrt{(1/12n)}}$$

x_0 is the time at which the test is truncated. If $U = 0$ then there is no trend and the failure rate is not changing. If $U < 0$ then the failure rate is decreasing and if $U > 0$ it is increasing.

This test could be applied to the analysis of software failures since they are an example of a continuous repair process.

Figure 6.4 Failure distribution of a large population

EXERCISES

1. Components, as described in the example of Section 6.2, are to be used in a system. It is required that these are preventively replaced such that there is only a 5% probability of their failing beforehand. After how many hours should each item be replaced?

2. A sample of 10 items is allowed to fail and the time for each failure is as follows:

    4, 6, 8, 11, 12, 13, 15, 17, 20, 21 (thousand hours)

Use the Weibull paper in this chapter to determine the reliability characteristic and the MTBF.

Part Three
Predicting Reliability and Risk

7 Essential reliability theory

7.1 WHY PREDICT RAMS?

Reliability prediction (i.e. modelling) is the process of calculating the anticipated system RAMS from assumed component failure rates. It provides a quantitative measure of how close a proposed design comes to meeting the design objectives and allows comparisons to be made between different design proposals. It has already been emphasized that reliability prediction is an imprecise calculation, but it is nevertheless a valuable exercise for the following reasons:

- It provides an early indication of a system's potential to meet the design reliability requirements.
- It enables an assessment of life cycle costs to be carried out.
- It enables one to establish which components, or areas, in a design contribute to the major portion of the unreliability.
- It enables trade-offs to be made as, for example, between reliability, maintainability and proof-test intervals in achieving a given availability.
- Its use is increasingly called for in invitations to tender, contracts and in safety–integrity standards.

It must be stressed that prediction is a design tool and not a precise measure of reliability. The main value of a prediction is in showing the relative reliabilities of modules so that allocations can be made. Whatever the accuracy of the exercise, if one module is shown to have double the MTBF of another then, when calculating values for modules in order to achieve the desired system MTBF, the values allocated to the modules should be in the same ratio. Prediction also permits a reliability comparison between different design solutions. Again, the comparison is likely to be more accurate than the absolute values. The accuracy of the actual predicted value will depend on:

1. Relevance of the failure rate data and the chosen environmental multiplication factors;
2. Accuracy of the Mathematical Model;
3. The absence of gross over-stressing in operation.
4. Tolerance of the design to component parametric drift.

The greater the number of different component types involved, the more likely that individual over- and under-estimates will cancel each other out.

7.2 PROBABILITY THEORY

The following basic probability rules are sufficient for an understanding of the system modelling involved in reliability prediction.

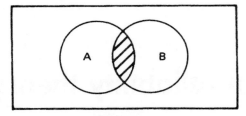

Figure 7.1

7.2.1 The Multiplication Rule

If two or more events can occur simultaneously, and their individual probabilities of occurring are known, then the probability of simultaneous events is the product of the individual probabilities. The shaded area in Figure 7.1 represents the probability of events A and B occurring simultaneously. Hence the probability of A and B occurring is:

$$Pab = Pa \times Pb$$

Generally

$$Pan = Pa \times Pb, \ldots\ldots\ldots, \times Pn$$

7.2.2 The Addition Rule

It is also required to calculate the probability of either event A *or* event B *or both* occurring. This is the area of the two circles in Figure 7.1. This probability is:

$$P(a \text{ or } b) = Pa + Pb - PaPb$$

being the sum of *Pa* and *Pb* less the area *PaPb* which is included twice. This becomes:

$$P(a \text{ or } b) = 1 - (1 - Pa)(1 - Pb)$$

Hence the probability of one or more of *n* events occurring is:

$$= 1 - (1 - Pa)(1 - Pb), \ldots, (1 - Pn)$$

7.2.3 The Binomial Theorem

The above two rules are combined in the Binomial Theorem. Consider the following example involving a pack of 52 playing cards. A card is removed at random, its suit noted, and then replaced. A second card is then removed and its suit noted. The possible outcomes are:

Two hearts
One heart and one other card
Two other cards

If p is the probability of drawing a heart then, from the multiplication rule, the outcomes of the experiment can be calculated as follows:

Probability of 2 hearts p^2
Probability of 1 heart $2pq$
Probability of 0 hearts q^2

Similar reasoning for an experiment involving 3 cards will yield:

Probability of 3 hearts p^3
Probability of 2 hearts $3p^2q$
Probability of 1 heart $3pq^2$
Probability of 0 hearts q^3

The above probabilities are the terms of the expressions $(p + q)^2$ and $(p + q)^3$. This leads to the general statement that if p is the probability of some random event, and if $q = 1 - p$, then the probabilities of 0, 1, 2, 3, . . ., outcomes of that event in n trials are given by the terms of the expansion:

$(p + q)^n$ which equals

$$p^n, np^{(n-1)}q, \frac{n(n-1)p^{(n-2)}q^2}{2!}, \ldots, q^n$$

This is known as the binomial expansion.

7.2.4 The Bayes Theorem

The marginal probability of an event is its simple probability. Consider a box of seven cubes and three spheres in which case the marginal probability of drawing a cube is 0.7. To introduce the concept of a Conditional Probability assume that four of the cubes are black and three white and that, of the spheres, two are black and one is white, as shown in Figure 7.2.

The probability of drawing a black article, given that it turns out to be a cube, is a conditional probability of 4/7 and ignores the possibility of drawing a sphere. Similarly the probability of drawing a black article, given that it turns out to be a sphere, is 2/3. On the other hand, the probability of drawing a black sphere is a Joint Probability. It acknowledges the possibility of drawing cubes and spheres and is therefore 2/10.

Figure 7.2

Comparing joint and conditional probabilities, the conditional probability of drawing a black article given that it is a sphere is the joint probability of drawing a black sphere (2/10) divided by the probability of drawing any sphere (3/10). The result is hence $\frac{2}{3}$. Therefore:

$$P_{b/s} = \frac{P_{bs}}{P_s} \text{ given that:}$$

$P_{b/s}$ is the conditional probability of drawing a black article given that it is a sphere; P_s is the simple or marginal probability of drawing a sphere; P_{bs} is the joint probability of drawing an article which is both black and a sphere.

This is known as the Bayes Theorem. It follows then that $P_{bs} = P_{b/s} . P_s$ or $P_{s/b} . P_b$. Consider now the probability of drawing a black sphere (P_{bs}) and the probability of drawing a white sphere (P_{ws}):

$$P_s = P_{bs} + P_{ws}$$

Therefore

$$P_s = P_{s/b} . P_b + P_{s/w} . P_w$$

and, in general,

$$P_x = P_{x/a} . P_a + P_{x/b} . P_b, \ldots, + P_{x/n} . P_n$$

which is the form applicable to prediction formulae.

7.3 RELIABILITY OF SERIES SYSTEMS

Consider the two valves connected in series which were described in Chapter 2.
One of the failure modes discussed was loss of supply which occurs if *either* valve fails closed. This situation, where any failure causes the system to fail, is known as *series* reliability. This must not be confused with the series configuration of the valves shown in Figure 2.1 It so happens that, for this loss of supply failure mode, the physical series and the reliability series diagrams coincide. When we consider the over-pressure case in the next section it will be seen that, although the valves are still in series, the reliability block diagram changes.

For loss of supply then, the reliability of the system is the probability that Valve A does not fail *and* Valve B does not fail.

From the multiplication rule in Section 7.2.1 then:

$$R_{ab} = R_a . R_b \text{ and, in general,}$$
$$R_{an} = R_a . R_b, \ldots, R_n$$

In the constant failure rate case where:

$$R_a = e^{-\lambda_a t}$$

Then

$$R_n = \exp\left[-(\lambda_a + \lambda_b, \ldots, \lambda_n)t\right]$$

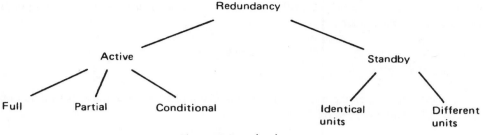

Figure 7.3 Redundancy

from which it can be seen that the system is also a constant failure rate unit whose reliability is of the form e^{-Kt}, where K is the sum of the individual failure rates. Provided that the two assumptions of constant failure rate and series modelling apply, then it is valid to speak of a system failure rate computed from the sum of the individual unit or component failure rates.

The practice of adding up the failure rates in a Component Count type prediction assumes that any single failure causes a system failure. It is therefore a worst-case prediction since, clearly, a Failure Mode Analysis against a specific failure mode will involve only those components which contribute to that top event.

Returning to the example of the two valves, assume that each has a failure rate of 7×10^{-6} per hour for the fail closed mode and consider the reliability for one year. One year has 8760 hours.

From the above:

$$\lambda_{\text{system}} = \lambda_a + \lambda_b = 14 \times 10^{-6} \text{ per hour}$$

$$\lambda t = 8760 \times 14 \times 10^{-6} = 0.1226$$

$$R_{\text{system}} = e^{-\lambda t} = 0.885$$

7.4 REDUNDANCY RULES

7.4.1 General types of redundant configuration

There are a number of ways in which redundancy can be applied. These are shown in diagram form in Figure 7.3. So far, we have met only the particular case of Full Active Redundancy. The models for the other cases will be described in the following sections. At present, we are considering redundancy without repair and it is assumed that failed redundant units remain failed until the whole system fails. The point concerning variable failure rate applies to each of the models.

7.4.2 Full active redundancy (without repair)

Continuing with our two-valve example, consider the over-pressure failure mode described in Chapter 2. There is no longer a reliability *series* situation since both valves need to fail open in order for the top event to occur. In this case a *parallel* reliability block diagram applies. Since either, or both, valves operating correctly is sufficient for system success, then the addition rule in Section 7.2.2 applies. For the two valves it is:

$$R_{\text{system}} = 1 - (1 - R_a)(1 - R_b) \text{ or, in another form}$$
$$R_{\text{system}} = R_a + R_b - R_a R_b$$

In other words, one minus the product of their unreliabilities. Let us assume that the fail open failure rate of a valve is 3×10^{-6} per hour:

$$R_a = R_b = e^{-\lambda t} \quad \text{where } \lambda t = 3 \times 10^{-6} \times 8760 = 0.026$$

$$e^{-\lambda t} = 0.974$$

$$R_{\text{system}} = 1 - (0.026)^2 = \underline{0.999}$$

If there were N items in this redundant configuration such that all may fail except one, then the expression becomes

$$R_{\text{system}} = 1 - (1 - R_a)(1 - R_b), \ldots, (1 - R_n)$$

There is a pitfall at this point which it is important to emphasize. The reliability of the system, after substitution of $R = e^{-\lambda t}$, becomes:

$$R_S = 2e^{-\lambda t} - e^{-2\lambda t}$$

It is very important to note that, unlike the series case, this combination of constant failure rate units exhibits a reliability characteristic which is *not* of the form e^{-Kt}. In other words, although constant failure rate units are involved, the failure rate of the system is variable. The MTBF can therefore be obtained only from the integral of reliability. In Chapter 2 we saw that

$$\text{MTBF} = \int_0^\infty R(t) \, dt$$

Hence

$$\text{MTBF} = \int_0^\infty (2e^{-\lambda t} - e^{-2\lambda t})$$

$$= 2/\lambda - 1/2\lambda$$

$$= 3/2\lambda$$

$$= \underline{3\theta/2} \text{ where } \theta \text{ is the MBTF of a single unit.}$$

In the above working we substituted θ for $1/\lambda$ which was correct because a unit was being considered for which constant λ applies. The danger now is to assume that the failure rate of the system is $2\lambda/3$. This is not true since the practice of inverting MTBF to obtain failure rate, and vice versa, is valid only for *constant failure rate*.

Figure 7.4 compares reliability against time, and failure rate against time, for series and redundant cases. As can be seen, the failure rate, initially zero, increases asymptotically. Reliability, in a redundant configuration, stays higher than for constant failure rate at the beginning but eventually falls more sharply. The greater the number of redundant units, the

Figure 7.4 Effect of redundancy on reliability and failure rate

longer the period of higher reliability and the sharper the decline. These features of redundancy apply, in principle, to all redundant configurations – only the specific values change.

7.4.3 Partial active redundancy (without repair)

Consider three identical units each with reliability R. Let $R + Q = 1$ so that Q is the unreliability (probability of failure in a given time). The binomial expression $(R + Q)^3$ yields the following terms:

$R^3, 3R^2Q, 3RQ^2, Q^3$ which are

$R^3, 3R^2(1 - R), 3R(1 - R)^2, (1 - R)^3$

This conveniently describes the probabilities of

0 , 1 , 2 , 3   failures of a single unit.

In Section 7.4.2 the reliability for full redundancy was seen to be:

$1 - (1 - R)^3$

This is consistent with the above since it can be seen to be 1 minus the last term. Since the sum of the terms is unity reliability it is therefore the sum of the first three terms which, being the probability of 0, 1 or 2 failures, is the reliability of a fully redundant system.

In many cases of redundancy, however, the number of units permitted to fail before system failure occurs is less than in full redundancy. In the example of three units full redundancy requires only one to function, whereas partial redundancy would exist if two units were required with only one allowed to fail. Once again the reliability can be obtained from the binomial expression since it is the probability of 0 or 1 failures which is given by the sum of the first two terms. Hence:

$$R_{\text{system}} = R^3 + 3R^2(1 - R)$$
$$= 3R^2 - 2R^3$$

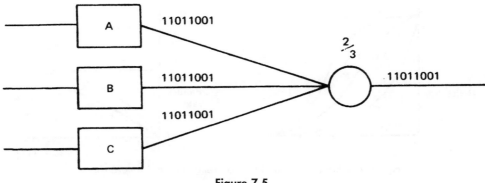

Figure 7.5

In general, if r items may fail out of n then the reliability is given as the sum of the first $r +$ 1 terms of the binomial expansion $(R + Q)^n$. Therefore

$$R = R^n + nR^{n-1}(1 - R) + \frac{n(n - 1)R^{n-2}(1 - R)^2}{2!} + \ldots$$

$$\ldots + \frac{n(n - 1) \ldots (n - r + 1)R^{n-r}(1 - R)^r}{r!}$$

7.4.4 Conditional active redundancy

This is best considered by an example. Consider the configuration in Figure 7.5. Three identical digital processing units (A, B and C) have reliability R. They are triplicated to provide redundancy in the event of failure and their identical outputs are fed to a two out of three majority voting gate. If two identical signals are received by the gate they are reproduced at the output. Assume that the voting gate is sufficiently more reliable than the units so that its probability of failure can be disregarded. Assume also that the individual units can fail either to an open circuit or a short circuit output. Random data bit errors are not included in the definition of system failure for the purpose of this example. The question arises as to whether the system has:

Partial Redundancy   1 unit may fail but no more, or
Full Redundancy   2 units may fail.

The answer is conditional on the mode of failure. If two units fail in a like mode (both outputs logic 1 or logic 0) then the output of the voting gate will be held at the same value and the system will have failed. If, on the other hand, they fail in unlike modes then the remaining unit will produce a correct output from the gate since it always sees an identical binary bit from one of the other units. This conditional situation requires the Bayes theorem introduced in Section 7.2.4. The equation becomes:

$$R_{\text{system}} = R_{\text{given} A} \cdot P_A + R_{\text{given} B} \cdot P_B, \ldots, + R_{\text{given} N} \cdot P_N$$

where A to N are mutually exclusive and $\sum_{i=A}^{i=N} P_i = 1$

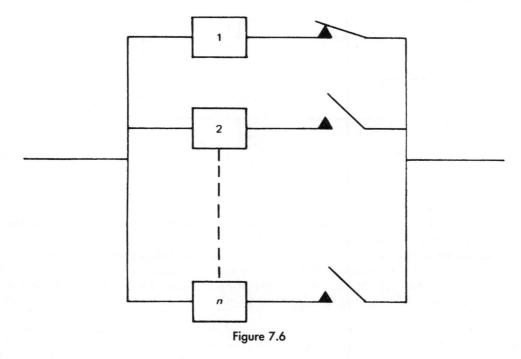

Figure 7.6

In this case the solution is:

$R_{\text{system}} = R_{\text{system given that in the event of failure 2 units fail alike}} \times P_{\text{failing alike}}$

$+$

$+R_{\text{system given that in the event of failure 2 units fail unalike}} \times P_{\text{failing unalike}}$

Therefore:

$$R_s = [R^3 + 3R^2(1 - R)] . P_A + [1 - (1 - R)^3] . P_B$$

since if two units fail alike there is partial redundancy and if two units fail unalike there is full redundancy. Assume that the probability of both failure modes is the same and that $P_A = P_B = 0.5$. The system reliability is therefore:

$$R_s = \frac{R^3 + 3R^2 - 3R^3 + 1 - 1 + 3R - 3R^2 + R^3}{2} = \frac{3R - R^3}{2}$$

7.4.5 Standby redundancy

So far, only active redundancy has been considered where every unit is operating and the system can function despite the loss of one or more units. Standby redundancy involves additional units which are activated only when the operating unit fails. A greater improvement, per added unit, is anticipated than with active redundancy since the standby units operate for less time. Figure 7.6 shows n identical units with item 1 active. Should a failure be detected then item 2 will be switched in its place. Initially, the following assumptions are made:

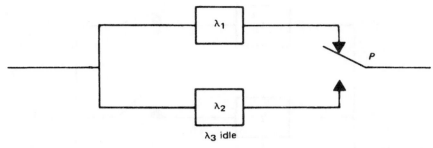

Figure 7.7

1. The means of sensing that a failure has occurred and for switching from the defective to the standby unit is assumed to be failure free.
2. The standby unit(s) are assumed to have identical, constant failure rates to the main unit.
3. The standby units are assumed not to fail while in the idle state.
4. As with the earlier calculation of active redundancy, defective units are assumed to remain so. No repair is effected until the system has failed.

Calculations involving redundancy and repair are covered in the next chapter. The reliability is then given by the first n terms of the Poisson expression:

$$R_{\text{system}} = R(t) = e^{-\lambda t}\left(1 + \lambda t + \frac{\lambda^2 t^2}{2!} \cdots \frac{\lambda^{(n-1)} t^{(n-1)}}{(n-1)!}\right)$$

which reduces, for two units to:

$$R_{\text{system}} = e^{-\lambda t}(1 + \lambda t)$$

Figure 7.7 shows the more general case of two units with some of the above assumptions removed. In the figure:

 λ_1 is the constant failure rate of the main unit,
 λ_2 is the constant failure rate of the standby unit when in use,
 λ_3 is the constant failure rate of the standby unit in the idle state,
 P is the one-shot probability of the switch performing when required.

The reliability is given by:

$$R_{\text{system}} = e^{-\lambda_1 t} + \frac{P\lambda_1}{\lambda_2 - \lambda_1 - \lambda_3}(e^{-(\lambda_1 + \lambda_3)t} - e^{-\lambda_2 t})$$

It remains only to consider the following failure possibilities. Let λ_4, λ_5 and λ_6 be the failure rates associated with the sums of the following failure modes:

For λ_4 – Dormant failures which inhibit failure sensing or changeover;
For λ_5 – Failures causing the incorrect switching back to the failed unit;
For λ_6 – False sensing of non-existent failure.

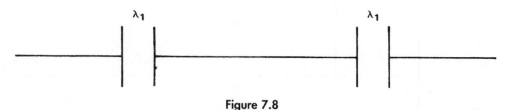

Figure 7.8

If we think about each of these in turn it will be seen that, from the point of view of the above model:

λ_4 is part of λ_3
λ_5 is part of λ_2
λ_6 is part of λ_1

In the analysis they should therefore be included in the appropriate category.

7.4.6 Load sharing

The following situation can be deceptive since, at first sight, it appears as active redundancy. Figure 7.8 shows two capacitors connected in series. Given that both must fail short circuit in order for the system to fail, we require a model for the system. It is *not* two units in active redundant configuration because if the first capacitor should fail (short circuit) then the voltage applied to the remaining one will be doubled and its failure rate greatly increased. This situation is known as load sharing and is mathematically identical to a standby arrangement.

Figure 7.9 shows two units in standby configuration. The switchover is assumed to be perfect (which is appropriate) and the standby unit has an idle failure rate equal to zero with a different (larger) failure rate after switchover. The main unit has a failure rate of twice the single capacitor.

7.5 GENERAL FEATURES OF REDUNDANCY

7.5.1 Incremental improvement

As was seen in Figure 7.4, the improvement resulting from redundancy is not spread evenly along the time axis. Since the MTBF is an overall measure obtained by integrating reliability

Figure 7.9

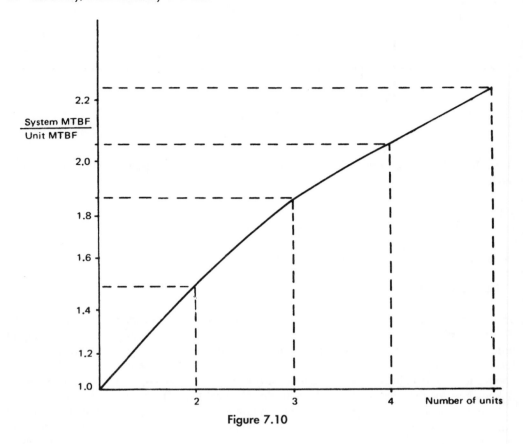

Figure 7.10

from zero to infinity, it is actually the area under the curve of reliability against time. For short missions (less than one MTBF in duration) the actual improvement in reliability is greater than would be suggested simply by comparing MTBFs. For this reason, the length of mission should be taken into account when evaluating redundancy.

As we saw in Section 7.4, the effect of duplicating a unit by active redundancy is to improve the MTBF by only 50%. This improvement falls off as the number of redundant units increases, as is shown in Figure 7.10. The effect is similar for other redundant configurations such as conditional and standby. Beyond a few units the improvement may even be offset by the unreliability introduced as a result of additional switching and other common mode effects dealt with in Section 8.2.

Figure 7.10 is not a continuous curve since only the points for integral numbers of units exist. It has been drawn, however, merely to illustrate the diminishing enhancement in MTBF as the number of units is increased.

7.5.2 Further comparisons of redundancy

Figure 7.11 shows two alternative configurations involving 4 units in active redundancy: (i) protects against short circuit failures whereas (ii) protects against short- and open-circuit conditions. As can be seen from Figure 7.12, (ii) has the higher reliability but is harder to

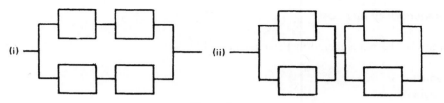

Figure 7.11

implement. If readers care to calculate the MTBF of (i), they will find that it can be less than for a single unit and, as can be seen from the curves, the area under the reliability curve (MTBF) is less. It is of value only for conditions where the short-circuit failure mode is more likely.

Figure 7.13 gives a comparison between units in both standby and active redundancy. For the simple model assuming perfect switching the standby configuration has the higher reliability, although, in practice, the associated hardware for sensing and switching will erode the advantage. On the other hand, it is not always easy to achieve active redundancy with true independence between units. In other words, the failure of one unit may cause or at least hasten the failure of another. This common mode effect will be explained in the next chapter (Section 8.2).

Figure 7.12

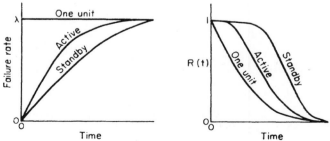

Figure 7.13

7.5.3 Redundancy and cost

It must always be remembered that redundancy adds:

> Capital cost
> Weight
> Spares
> Space
> Preventive maintenance
> Power consumption
> Failures at the unit level (hence more corrective maintenance)

Each of these contributes substantially to cost.

EXERCISES

1. Calculate the MTBF of the system shown in the following block diagram.

2. The following block diagram shows a system whereby unit B may operate with units D or E but where unit A may only operate with unit D, or C with E. Derive the reliability expression.

8 Methods of modelling

In Chapter 1 (Section 1.3) and Chapter 7 (Section 7.1) the limitations of reliability of prediction were emphasized. This chapter describes, in some detail, the available methods.

8.1 BLOCK DIAGRAM AND MARKOV ANALYSIS

8.1.1 Reliability block diagrams

The following is the general approach to block diagram analysis.

Establish failure criteria

Define what constitutes a system failure since this will determine which failure modes at the component level actually cause a system to fail. There may well be more than one type of system failure, in which case a number of predictions giving different reliabilities will be required. This step is absolutely essential if the predictions are to have any significance. It was explained, in Section 2.1, how different system failure modes can involve quite different component failure modes and, indeed, even different series/redundant configurations.

Establish a reliability block diagram

It is necessary to describe the system as a number of functional blocks which are interconnected according to the effect of each block failure on the overall system reliability.

Figure 8.1 is a series diagram representing a system of two blocks such that the failure of either block prevents operation of the system. Figure 8.2 shows the situation where both blocks must fail in order for the system to fail. This is known as a parallel, or redundancy, case. Figure

Figure 8.1

Figure 8.2

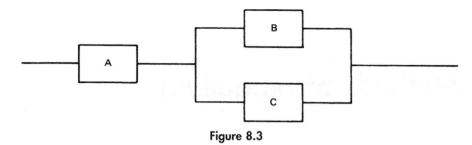

Figure 8.3

8.3, shows a combination of series and parallel reliability. It represents a system which will fail if block A fails or if both block B and block C fail. The failure of B or C alone is insufficient to cause system failure.

A number of general rules should be borne in mind when defining the blocks.

1. Each block should represent the maximum number of components in order to simplify the diagram.
2. The function of each block should be easily identified.
3. Blocks should be mutually independent in that failure in one should not affect the probability of failure in another (see Section 8.2).
4. Blocks should not contain any significant redundancy otherwise the addition of failure rates, within the block, would not be valid.
5. Each replaceable unit should be a whole number of blocks.
6. Each block should contain one technology, that is, electronic or electro-mechanical.
7. There should be only one environment within a block.

Failure mode analysis

Failure Mode and Effect Analysis (FMEA) is described later in Chapter 9 (Section 9.3). It provides block failure rates by examining individual component failure modes and failure rates. Given a constant failure rate and no internal redundancy, each block will have a failure rate predicted from the sum of the failure rates on the FMEA worksheet.

Calculation of system reliability

Relating the block failure rates to the system reliability is a question of mathematical modelling which is the subject of the rest of this section. In the event that the system reliability prediction fails to meet the objective then improved failure rate (or down time) objectives must be assigned to each block by means of reliability allocation.

Reliability allocation

The importance of reliability allocation is stressed in Chapter 11 and an example is calculated. The block failure rates are taken as a measure of the complexity, and improved, suitably weighted, objectives are set.

8.1.2 The Markov model for repairable systems

In Chapter 7 the basic rules for series and redundant systems were explained. For redundant systems, however, the equations only catered for the case of redundancy with no repair of failed units. In other words, the reliability would be the probability of the system not failing given that any failed redundant units stayed failed.

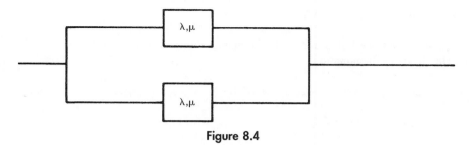

Figure 8.4

In order to cope with systems whose redundant units are subject to a repair strategy the Markov analysis technique is used. The technique assumes both constant failure rate and constant repair rate. For other distributions (e.g. Weibull failure rate process or log Normal repair times) Monte Carlo simulation methods are more appropriate (see Section 9.5).

The Markov method for calculating the MTTF of a system with repair is to consider the 'states' in which the system can exist. Figure 8.4 shows a system with two identical units each having failure rate λ and repair rate (reciprocal of mean down time) μ. The system can be in each of three possible states.

State (0)   Both units operating
State (1)   One unit operating, the other having failed
State (2)   Both units failed

It is important to remember one rule with Markov analysis, namely, that the probabilities of changing state are dependent only on the state itself. In other words, the probability of failure or of repair is not dependent on the past history of the system.

Let $P_i(t)$ be the probability that the system is in state (i) at time t and assume that the initial state is (0).

Therefore

$$P_0(0) = 1 \text{ and } P_1(0) = P_2(0) = 0$$

Therefore

$$P_0(t) + P_1(t) + P_2(t) = 1$$

We shall now calculate the probability of the system being in each of the three states at time $t + \Delta t$. The system will be in state (0) at time $t + \Delta t$ if:

1. The system was in state (0) at time t and no failure occurred in either unit during the interval Δt, or,
2. The system was in state (1) at time t, no further failure occurred during Δt, and the failed unit was repaired during Δt.

The probability of only one failure occurring in one unit during that interval is simply $\lambda \Delta t$ (valid if Δt is small, which it is). Consequently $(1 - \lambda \Delta t)$ is the probability that no failure will occur in one unit during the interval. The probability that both units will be failure free during the interval is, therefore,

$$(1 - \lambda\Delta t)(1 - \lambda\Delta t) \quad 1 - 2\lambda\Delta t$$

The probability that one failed unit will be repaired within Δt is $\mu\Delta t$, provided that Δt is very small. This leads to the equation:

$$P_0(t + \Delta t) = [P_0(t) \times (1 - 2\lambda\Delta t)] + [P_1(t) \times (1 - \lambda\Delta t) \times \mu\Delta t]$$

Similarly, for states 1 and 2:

$$P_1(t + \Delta t) = [P_0(t) \times 2\lambda\Delta t] + [P_1(t) \times (1 - \lambda\Delta t) \times (1 - \mu\Delta t)]$$

$$P_2(t + \Delta t) = [P_1(t) \times \lambda\Delta t] + P_2(t)$$

Now the limit as Δt 0 of $[P_i(t + \Delta t) - P_i(t)]/\Delta t$ is $P_i(t)$ and so the above yield:

$$\dot{P}_0(t) = -2\lambda P_0(t) + \mu P_1(t)$$

$$\dot{P}_1(t) = 2\lambda P_0(t) - (\lambda + \mu)P_1(t)$$

$$\dot{P}_2(t) = P_1(t)\lambda$$

In matrix notation this becomes:

$$
\begin{vmatrix} \dot{P}_0 \\ \dot{P}_1 \\ \dot{P}_2 \end{vmatrix}
=
\begin{vmatrix} -2\lambda & \mu & 0 \\ 2\lambda & -(\lambda + \mu) & 0 \\ 0 & \lambda & 0 \end{vmatrix}
\begin{vmatrix} P_0 \\ P_1 \\ P_2 \end{vmatrix}
$$

The elements of this matrix can also be obtained by means of a Transition Diagram. Since only one event can take place during a small interval, Δt, the transitions between states involving only one repair or one failure are considered. Consequently, the transitions (with transition rates) are:

by failure of either unit

by failure of the remaining active unit,

by repair of the failed unit of state 1.

The transition diagram is:

Finally closed loops are drawn at states 0 and 1 to account for the probability of not changing state. The rates are easily calculated as minus the algebraic sum of the rates associated with the lines leaving that state. Hence:

A (3×3) matrix, $(a_{i,j})$, can now be constructed, where $i = 1, 2, 3; j = 1, 2, 3; a_{i,j}$ is the character on the flow line pointing from state j to state i. If no flow line exists the corresponding matrix element is zero. We therefore find the same matrix as before.

The MTTF is defined as

$$\theta_s = \int_0^\infty R(t) \, dt$$

$$= \int_0^\infty [P_0(t) + P_1(t)] \, dt$$

$$= \int_0^\infty P_0(t) \, dt + \int_0^\infty P_1(t) \, dt$$

$$= T_0 + T_1$$

The values of T_0 and T_1 can be found by solving the following:

$$\int_0^\infty \begin{vmatrix} \dot{P}_0(t) \\ \dot{P}_1(t) \\ \dot{P}_2(t) \end{vmatrix} dt = \int_0^\infty \begin{vmatrix} -2\lambda & \mu & 0 \\ 2\lambda & -(\lambda + \mu) & 0 \\ 0 & \lambda & 0 \end{vmatrix} \begin{vmatrix} P_0 \\ P_1 \\ P_2 \end{vmatrix} dt$$

Since the (3×3) matrix is constant we may write

$$\int_0^\infty \begin{vmatrix} \dot{P}_0(t) \\ \dot{P}_1(t) \\ \dot{P}_2(t) \end{vmatrix} dt = \begin{vmatrix} -2\lambda & \mu & 0 \\ 2\lambda & -(\lambda + \mu) & 0 \\ 0 & \lambda & 0 \end{vmatrix} \int_0^\infty \begin{vmatrix} P_0 \\ P_1 \\ P_2 \end{vmatrix} dt$$

or

$$\begin{vmatrix} \int_0^\infty \dot{P}_0(t) \, dt \\ \int_0^\infty \dot{P}_1(t) \, dt \\ \int_0^\infty \dot{P}_2(t) \, dt \end{vmatrix} = \begin{vmatrix} -2\lambda & \mu & 0 \\ 2\lambda & -(\lambda + \mu) & 0 \\ 0 & \lambda & 0 \end{vmatrix} \begin{vmatrix} \int_0^\infty P_0(t) \, dt \\ \int_0^\infty P_1(t) \, dt \\ \int_0^\infty P_2(t) \, dt \end{vmatrix}$$

or

$$\begin{vmatrix} P_0(\infty) - P_0(0) \\ P_1(\infty) - P_1(0) \\ P_2(\infty) - P_2(0) \end{vmatrix} = \begin{vmatrix} -2\lambda & \mu & 0 \\ 2\lambda & -(\lambda + \mu) & 0 \\ 0 & \lambda & 0 \end{vmatrix} \begin{vmatrix} T_0 \\ T_1 \\ T_2 \end{vmatrix}$$

Taking account of

$$P_0(0) = 1; \quad P_1(0) = P_2(0) = 0$$
$$P_0(\infty) = P_1(\infty) = 0; \quad P_2(\infty) = 1$$

we may reduce the equation to

$$
\begin{vmatrix} -1 \\ 0 \\ 1 \end{vmatrix}
=
\begin{vmatrix} -2\lambda & \mu & 0 \\ 2\lambda & -(\lambda + \mu) & 0 \\ 0 & \lambda & 0 \end{vmatrix}
\begin{matrix} T_0 \\ T_1 \\ T_2 \end{matrix}
$$

or

$$-1 = -2\lambda T_0 + \mu T_1$$
$$0 = 2\lambda T_0 - (\lambda + \mu)T_1$$
$$1 = \lambda T_1$$

Solving this set of equations

$$T_0 = \frac{\lambda + \mu}{2\lambda^2} \quad \text{and} \quad T_1 \frac{1}{\lambda}$$

so that

$$\theta_s = T_0 + T_1 = \frac{1}{\lambda} + \frac{\lambda + \mu}{2\lambda^2} = \frac{3\lambda + \mu}{2\lambda^2}$$

that is,

$$\theta_s = \frac{3\lambda + \mu}{2\lambda^2}$$

The Markov analysis technique can equally well be applied to calculating the steady state unavailability. To do this, one must consider recovery from the system failed state. The transition diagram is therefore modified to allow for this and takes the form:

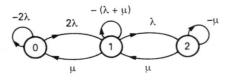

The path from state (2) to state (1) has a rate of μ. This reflects the fact that only one repair can be considered at one time. If more resources were available then two simultaneous repairs could be conducted and the rate would become 2μ. Constructing a matrix as shown earlier:

$$
\begin{vmatrix} \dot{P_0} \\ \dot{P_1} \\ \dot{P_2} \end{vmatrix} = \begin{vmatrix} -2\lambda & \mu & 0 \\ 2\lambda & -(\lambda + \mu) & \mu \\ 0 & \lambda & -\mu \end{vmatrix} \begin{vmatrix} P_0 \\ P_1 \\ P_2 \end{vmatrix}
$$

Since the steady state is being modelled the rate of change of probability of being in a particular state is zero, hence:

$$
\begin{vmatrix} -2\lambda & \mu & 0 \\ 2\lambda & -(\lambda + \mu) & \mu \\ 0 & \lambda & -\mu \end{vmatrix} \begin{vmatrix} P_0 \\ P_1 \\ P_2 \end{vmatrix} = \begin{vmatrix} 0 \\ 0 \\ 0 \end{vmatrix}
$$

Therefore

$$
\begin{aligned}
-2\lambda P_0 + \mu P_1 &= 0 \\
2\lambda P_0 - (\lambda + \mu)P_1 + \mu P_2 &= 0 \\
\lambda P_1 - \mu P_2 &= 0
\end{aligned}
$$

However, the probability of being in one of the states is unity, therefore

$$
P_0 + P_1 + P_2 = 1
$$

The system unavailability is required and this is represented by P_2, namely, the probability of being in the failed state. Thus:

$$
\text{Unavailability} = P_2 = 2\lambda^2/(2\lambda^2 + \mu^2 + 2\lambda\mu)
$$

The effect of the spares quantity on these models is dealt with in Chapter 16.

8.1.3 A summary of Markov results (revealed failures)

In the previous section the case for simple active redundancy was explained. The following tables provide the results and approximations for a range of redundancy and repair cases:

1. Active redundancy – two identical units.

$$= \frac{3\lambda + \mu}{2\lambda^2} \cong \frac{\mu}{2\lambda^2} \quad \text{if} \quad \mu \gg \lambda$$

2. Full active redundancy – n identical units with n repair crews.

$$\theta = \frac{\mu^{(n-1)}}{n\lambda^n} \qquad \text{if} \quad \mu \gg \lambda$$

3. Standby redundancy – two identical units.

$$\theta = \frac{2\lambda + \mu}{\lambda^2} \cong \frac{\mu}{\lambda^2} \quad \text{if} \quad \mu \gg \lambda$$

4. Standby redundancy – n identical units with n repair crews.

$$\theta = \frac{\mu^{(n-1)}}{\lambda^n} \qquad \text{if} \quad \mu \gg \lambda$$

5. Active redundancy – two different units

$$\theta = \frac{(\lambda_a + \mu_b)(\lambda_b + \mu_a) + \lambda_a(\lambda_a + \mu_b) + \lambda_b(\lambda_b + \mu_a)}{\lambda_a \lambda_b (\lambda_a + \lambda_b + \mu_a + \mu_b)}$$

6. Partial active redundancy with n repair crews

Table 8.1 System MTTF table (n crews)

Total number of units		1	2	3	4
	1	$\dfrac{1}{\lambda}$			
	2	$\dfrac{3\lambda + \mu}{2\lambda^2}$	$\dfrac{1}{2\lambda}$		
	3	$\dfrac{11\lambda^2 + 7\lambda\mu + 2\mu^2}{6\lambda^3}$	$\dfrac{5\lambda + \mu}{6\lambda^2}$	$\dfrac{1}{3\lambda}$	
	4	$\dfrac{25\lambda^3 + 23\lambda^2\mu + 13\lambda\mu^2 + 3\mu^3}{12\lambda^4}$	$\dfrac{13\lambda^2 + 5\lambda\mu + \mu^2}{12\lambda^3}$	$\dfrac{7\lambda + \mu}{12\lambda^2}$	$\dfrac{1}{4\lambda}$

Number of units required to operate

7. Partial active redundancy with only one repair crew

Table 8.2 System MTTF table (1 crew)

Total number of units		1	2	3	4
	1	$1/\lambda$			
	2	$(3\lambda + \mu)/2\lambda^2$	$1/2\lambda$		
	3	$(11\lambda^2 + 4\lambda\mu + \mu^2)/6\lambda^3$	$(5\lambda + \mu)/6\lambda^2$	$1/3\lambda$	
	4	$\dfrac{50\lambda^3 + 18\lambda^2\mu + 5\lambda\mu^2 + \mu^3}{24\lambda^4}$	$\dfrac{26\lambda^2 + 6\lambda\mu + \mu^2}{24\lambda^3}$	$\dfrac{7\lambda + \mu}{12\lambda^2}$	$1/4\lambda$

Number of units required to operate

If $\mu > \lambda$ then Table 8.1 can be expressed as Table 8.3 and Table 8.2 as Table 8.4.

Table 8.3 System failure rates (n crews)

Total number of units		1	2	3	4
	1	λ			
	2	$2\lambda^2$ MDT	2λ		
	3	$3\lambda^3$ MDT^2	$6\lambda^2$ MDT	3λ	
	4	$4\lambda^4$ MDT^3	$12\lambda^3$ MDT^2	$12\lambda^2$ MDT	4λ

Number of units required to operate

Table 8.4 System failure rates (1 crew)

Total number of units		1	2	3	4
	1	λ			
	2	$2\lambda^2$ MDT	2λ		
	3	$6\lambda^3$ MDT^2	$6\lambda^2$ MDT	3λ	
	4	$24\lambda^4$ MDT^3	$24\lambda^3$ MDT^2	$12\lambda^2$ MDT	4λ

Number of units required to operate

Tables of system unavailabilities can now be formulated. For the case of n repair crews the system MDT is the unit MDT divided by the number of items needed to fail. Table 8.5 is thus obtained by multiplying the cells in Table 8.3 by MDT/(number to fail). For the case of a single repair crew the system MDT will be the same as for the unit MDTs. Thus, Table 8.6 is obtained by multiplying the cells in Table 8.4 by MDT.

Table 8.5 System unavailability (n crews)

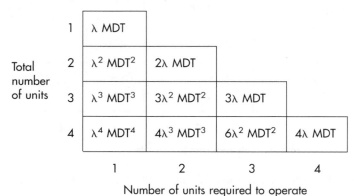

Total number of units		1	2	3	4
	1	$\lambda\,MDT$			
	2	$\lambda^2\,MDT^2$	$2\lambda\,MDT$		
	3	$\lambda^3\,MDT^3$	$3\lambda^2\,MDT^2$	$3\lambda\,MDT$	
	4	$\lambda^4\,MDT^4$	$4\lambda^3\,MDT^3$	$6\lambda^2\,MDT^2$	$4\lambda\,MDT$

Number of units required to operate

Table 8.6 System unavailability (1 crew)

Total number of units		1	2	3	4
	1	$\lambda\,MDT$			
	2	$2\lambda^2\,MDT^2$	$2\lambda\,MDT$		
	3	$6\lambda^3\,MDT^3$	$6\lambda^2\,MDT^2$	$3\lambda\,MDT$	
	4	$24\lambda^4\,MDT^4$	$24\lambda^3\,MDT^3$	$12\lambda^2\,MDT^2$	$4\lambda\,MDT$

Number of units required to operate

However, it is important to remember that the above 2 Tables were developed on the assumption that the SYSTEM MDT is the same as the UNIT MDT. This will not always be the case. Sometimes a failed system is a totally different scenario to that of repairing a failed unit. For example, a spurious (revealed) failure of an alarm will cause a particular repair activity. The failure of 2 alarms, on the other hand, may lead to a plant shutdown with quite different consequences and a different MDT. In that case Tables 8.3 and 8.4 must be multiplied by the SYSTEM MDT to obtain the Unavailabilities and Tables 8.5 and 8.6 would need to be modified.

8.1.4 Unrevealed failures

It is usually the case, with unattended equipment, that redundant items are not repaired immediately they fail. Either manual or automatic proof-testing takes place at regular intervals for the purpose of repairing or replacing failed redundant units. A system failure occurs when the redundancy is insufficient to sustain operation between visits. This is not as effective as immediate repair but costs considerably less in maintenance effort.

If the system is visited every T hours for the inspection of failed units (sometimes called the proof-test interval) then the Average Down Time AT THE UNIT LEVEL is

$$\frac{T}{2} + \text{Repair Time}$$

In general, given that the mean time to fail of a unit is much greater than the proof test interval then if z events need to occur for THE WHOLE SYSTEM to fail, these events will be randomly distributed between tests and, on average, the SYSTEM will be in a failed state for a time:

$$T/(z + 1)$$

For a system where r units are required to operate out of n then $n - r + 1$ must fail for the system to fail and so the SYSTEM down time becomes:

$$T/(n - r + 2) \qquad \text{(Equation 1)}$$

The probability of an individual UNIT failing between proof tests is simply:

$$\lambda T$$

For r out of n, the probability of the system failing prior to the next proof test is approximately the same as the probability of $n - r + 1$ units failing. This is:

$$\frac{n!}{(r - 1)! \, (n - r + 1)!} (\lambda T)^{\,n-r+1}$$

The failure rate is obtained by dividing this formula by T:

$$\frac{n!}{(r - 1)! \, (n - r + 1)!} \lambda^{n-r+1} \, T^{n-r} \qquad \text{(Equation 2)}$$

This yields Table 8.7.
Multiplying these failure rates by the system down time yields Table 8.8, i.e. (Equation 1) $\times$ (Equation 2).

Table 8.7 System failure rates

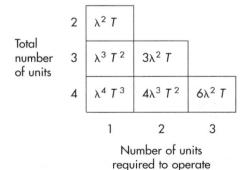

Total number of units	1	2	3
2	$\lambda^2 T$		
3	$\lambda^3 T^2$	$3\lambda^2 T$	
4	$\lambda^4 T^3$	$4\lambda^3 T^2$	$6\lambda^2 T$

Number of units required to operate

Table 8.8 System unavailability

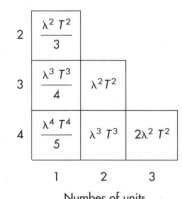

Total number of units	1	2	3
2	$\dfrac{\lambda^2 T^2}{3}$		
3	$\dfrac{\lambda^3 T^3}{4}$	$\lambda^2 T^2$	
4	$\dfrac{\lambda^4 T^4}{5}$	$\lambda^3 T^3$	$2\lambda^2 T^2$

Number of units required to operate

Once again it is important to realize that the MDT when the redundancy has been defeated, and the system has thus failed, has been assumed to be the same as for failed units. If this is not the case then the Unavailability must be obtained by multiplying Table 8.7 by the appropriate system MDT.

The above formulae apply to proof tests which are carried out simultaneously. If the proof-tests of redundant items are staggered (i.e. January/July) the Unavailability is improved (see the author's paper in *Journal of Safety & Reliability Society* (Autumn 2001).

8.2 COMMON CAUSE (DEPENDENT) FAILURE

8.2.1 What is CCF?

Common cause failures often dominate the unreliability of redundant systems by virtue of defeating the random coincident failure feature of redundant protection. Consider the duplicated system in Figure 8.5. The failure rate of the redundant element (in other words the coincident failures) can be calculated using the formula developed in Section 8.1, namely $2\lambda^2\text{MDT}$. Typical figures of 10 per million hours failure rate and 24 hours down time lead to a failure rate of 2 $\times 10^{-10} \times 24 = 0.0048$ per million hours. However, if only one failure in twenty is of such a nature as to affect both channels and thus defeat the redundancy, it is necessary to add the series element, λ_2, whose failure rate is $5\% \times 10^{-5} = 0.5$ per million hours. The effect is to swamp the redundant part of the prediction. This sensitivity of system failure to CCF places emphasis on the credibility of CCF estimation and thus justifies efforts to improve the models.

Whereas simple models of redundancy (developed in Section 8.1) assume that failures are both random and independent, common cause failure (CCF) modelling takes account of failures which are linked, due to some dependency, and therefore occur simultaneously or, at least, within a sufficiently short interval as to be perceived as simultaneous.

Two examples are:

(a) the presence of water vapour in gas causing both valves in twin streams to seize due to icing. In this case the interval between the two failures might be in the order of days. However, if the proof-test interval for this dormant failure is two weeks then the two failures will, to all intents and purposes, be simultaneous.
(b) inadequately rated rectifying diodes on identical twin printed circuit boards failing simultaneously due to a voltage transient.

Typically, causes arise from

(a) Requirements: Incomplete or conflicting
(b) Design: Common power supplies, software, emc, noise
(c) Manufacturing: Batch related component deficiencies
(d) Maintenance/Operations: Human induced or test equipment problems
(e) Environment: Temperature cycling, electrical interference, etc.

Defences against CCF involve design and operating features which form the assessment criteria shown in the next section.

The term common mode failure (CMF) is also frequently used and a brief explanation of the difference between CMF and CCF is therefore necessary. CMF refers to coincident failures of the same mode, in other words failures which have an identical appearance or effect. On the other hand, the term CCF implies that the failures have the same underlying cause. It is possible

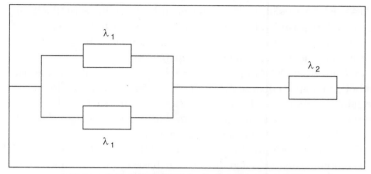

Figure 8.5 Reliability block diagrams for CCF

(although infrequent) for two CMFs not to have a common cause and, conversely, for two CCFs not to manifest themselves in the same mode. In practice the difference is slight and unlikely to affect the data, which rarely contain sufficient detail to justify any difference in the modelling. Since the models described in this section involve assessing defences against the CAUSES of coincident failure CCF will be used throughout.

8.2.2 Types of CCF model

Various approaches to modelling are:

(a) The simple BETA (β) model, which assumes that a fixed proportion (β) of the failures arise from a common cause. The estimation of (β) is assessed according to the system. (Note the Beta used in this context has no connection with the shape parameter used in the Weibull method, Chapter 6) The method is based on very limited historical data.

In Figure 8.5 (λ_1) is the failure rate of a single redundant unit and (λ_2) is the common cause failure rate such that (λ_2) = $\beta(\lambda_1)$ for the simple BETA model and also the Partial BETA model, in (b) below.

(b) The PARTIAL BETA model, also assumes that a fixed proportion of the failures arise from a common cause. It is more sophisticated than the simple BETA model in that the contributions to BETA are split into groups of design and operating features which are believed to influence the degree of CCF. Thus the BETA factor is made up by adding together the contributions from each of a number of factors within each group. In traditional Partial Beta models the following groups of factors, which represent defences against CCF, can be found:

- Similarity (Diversity between redundant units reduces CCF)
- Separation (Physical distance and barriers reduce CCF)
- Complexity (Simpler equipment is less prone to CCF)
- Analysis (Previous FMEA and field data analysis will have reduced CCF)
- Procedures (Control of modifications and of maintenance activities can reduce CCF)
- Training (Designers and maintainers can help to reduce CCF by understanding root causes)
- Control (Environmental controls can reduce susceptibility to CCF, e.g. weather proofing of duplicated instruments)
- Tests (Environmental tests can remove CCF prone features of the design, e.g. emc testing)

The PARTIAL BETA model is also represented by the reliability block diagram shown in Figure 8.5. BETA is assumed to be made up of a number of partial βs, each contributed to by the various groups of causes of CCF. β is then estimated by reviewing and scoring each of the contributing factors (e.g. diversity, separation).

(c) The System Cut-off model offers a single failure rate for all failures (independent and dependent both combined). It argues that the dependent failure rate dominates the coincident failures. Again, the choice is affected by system features such as diversity and separation. It is the least sophisticated of the models in that it does not base the estimate of system failure rate on the failure rate of the redundant units.

(d) The Boundary model uses two limits of failure rate. Namely, limit A which assumes all failures are common cause (λ_u) and limit B which assumes all failures are random (λ_1). The system failure rate is computed using a model of the following type:

$$\lambda = (\lambda_1^n \ \lambda_u)^{1/(n+1)}$$

where the value of n is chosen according to the degree of diversity between the redundant units. n is an integer, normally from 0–4, which increases with the level of diversity between redundant units. It is chosen in an arbitrary and subjective way. This method is a mathematical device, having no foundation in empirical data, which relies on a subjective assessment of the value of n. It provides no traceable link (as does the Partial BETA method) between the assessment of n and the perceived causes of CCF. Typical values of n for different types of system are:

CONFIGURATION	MODE OF OPERATION	PRECAUTIONS AGAINST CCF	n
Redundant equipment/system	Parallel	No segregation of services or supplies	0
Redundant equipment/system	Parallel	Full segregation of services or supplies	1
Redundant equipment/system	Duty/standby	No segregation of services or supplies	1
Redundant equipment/system	Duty/standby	Full segregation of services or supplies	2
Diverse equipment or system	Parallel	No segregation of services or supplies	2
Diverse equipment or system	Parallel	Full segregation of services or supplies	3
Diverse equipment or system	Duty/standby	No segregation of services or supplies	3
Diverse equipment or system	Duty/standby	Full segregation of services or supplies	4

(e) The Multiple Greek Letter model is similar to the BETA model but assumes that the BETA ratio varies according to the number of coincident failures. Thus two coincident failures and three coincident failures would have different BETA's. However, in view of the inaccuracy inherent in the approximate nature of these models it is considered to be too sophisticated and cannot therefore be supported by field data until more detailed information is available.

All the models are, in their nature, approximate but, because CCF failure rates (which are in the order of β λ) are much greater than the coincident independent failures (in the order of λ^n), then greater precision in estimating CCF is needed than for the redundant coincident models described in Section 8.1.

8.2.3 The BETAPLUS model

The BETAPLUS model has been developed from the Partial Beta method, by the author, because:

- it is objective and maximizes traceability in the estimation of BETA. In other words the choice of checklist scores when assessing the design, can be recorded and reviewed.
- it is possible for any user of the model to develop the checklists further to take account of any relevant failure causal factors that may be perceived.
- it is possible to calibrate the model against actual failure rates, albeit with very limited data.
- there is a credible relationship between the checklists and the system features being analysed. The method is thus likely to be acceptable to the non-specialist.
- the additive scoring method allows the partial contributors to β to be weighted separately.
- the β method acknowledges a direct relationship between (λ_2) and (λ_1) as depicted in Figure 8.5.
- it permits an assumed 'non-linearity' between the value of β and the scoring over the range of β.

The Partial BETA model includes the following enhancements:

(a) CATEGORIES OF FACTORS:
Whereas existing methods rely on a single subjective judgement of score in each category, the BETAPLUS method provides specific design and operational related questions to be answered in each category. Specific questions are individually scored, in each category (i.e. separation, diversity, complexity, assessment, procedures, competence, environmental control, environmental test) thereby permitting an assessment of the design and its operating and environmental factors. Other BETA methods only involve a single scoring of each category (e.g. a single subjective score for diversity).

(b) SCORING:
The maximum score for each question has been weighted by calibrating the results of assessments against known field operational data. Programmable and non-programmable equipment have been accorded slightly different checklists in order to reflect the equipment types (see Appendix 10).

(c) TAKING ACCOUNT OF DIAGNOSTIC COVERAGE:
Since CCF are not simultaneous, an increase in auto-test or proof-test frequency will reduce β since the failures may not occur at precisely the same moment. Thus, more frequent testing will prevent some CCF. Some defences will protect against the type of failure which increased proof-test might identify (for example failures in parallel channels where diversity would be beneficial). Other defences will protect against the type of failure which increased proof-test is unlikely to identify (for example failures prevented as a result of long term experience with the type of equipment) and this is reflected in the model.

(d) SUB-DIVIDING THE CHECKLISTS ACCORDING TO THE EFFECT OF DIAGNOSTICS:
Two columns are used for the checklist scores. Column (A) contains the scores for those features of CCF protection which are perceived as being enhanced by an increase of diagnostic frequency (either proof-test or auto-test). Column (B), however, contains the scores for those features thought not to be enhanced by an improvement in diagnostic frequency. In some cases the score has been split between the two columns, where it is thought that some, but not all, aspects of the feature are affected.

(e) ESTABLISHING A MODEL:
The model allows the scoring to be modified by the frequency and coverage of diagnostic test. The (A) column scores are modified by multiplying by a factor (C) derived from diagnostic related considerations. This (C) score is based on the diagnostic frequency and coverage. (C) is in the range 1 to 3. BETA is then estimated from the following RAW SCORE total:

$$S = RAW\ SCORE = (A\ \ C) + B$$

It is assumed that the effect of the diagnostic score (C) on the effectiveness of the (A) features is linear. In other words each failure mode is assumed to be equally likely to be revealed by the diagnostics. Only more detailed data can establish if this is not a valid assumption.

(f) NON-LINEARITY:
There are currently no CCF data to justify departing from the assumption that, as BETA decreases (i.e. improves) then successive improvements become proportionately harder to achieve. Thus the relationship of the BETA factor to the raw score [(A C) + B] is assumed to be exponential and this non-linearity is reflected in the equation which translates the raw score into a BETA factor.

(g) EQUIPMENT TYPE:
The scoring has been developed separately for programmable and non-programmable equipment, in order to reflect the slightly different criteria which apply to each type of equipment.

(i) CALIBRATION:
The model was calibrated against the author's field data.

Checklists and scoring of the (A) and (B) factors in the model.
    Scoring criteria were developed to cover each of the categories (i.e. separation, diversity, complexity, assessment, procedures, competence, environmental control, environmental test). Questions have been assembled to reflect the likely features which defend against CCF. The scores were then adjusted to take account of the relative contributions to CCF in each area, as shown in the author's data. The score values have been weighted to calibrate the model against the data.
    When addressing each question a score, less than the maximum of 100% may be entered. For example, in the first question, if the judgement is that only 50% of the cables are separated then 50% of the maximum scores (15 and 52) may be entered in each of the (A) and (B) columns (7.5 and 26).
    The checklists are presented in two forms (see Appendix 10) because the questions applicable to programmable based equipments will be slightly different to those necessary for non-programmable items (e.g. field devices and instrumentation).
    The headings (expanded with scores in Appendix 10) are:

(1) SEPARATION/SEGREGATION
(2) DIVERSITY/REDUNDANCY
(3) COMPLEXITY/DESIGN/APPLICATION/MATURITY/EXPERIENCE
(4) ASSESSMENT/ANALYSIS and FEEDBACK OF DATA
(5) PROCEDURES/HUMAN INTERFACE
(6) COMPETENCE/TRAINING/SAFETY CULTURE
(7) ENVIRONMENTAL CONTROL
(8) ENVIRONMENTAL TESTING

Assessment of the diagnostic interval factor (C).

In order to establish the (C) score it is necessary to address the effect of the frequency and coverage of proof-test or auto-test. The diagnostic coverage, expressed as a percentage, is an estimate of the proportion of failures which would be detected by the proof-test or auto-test. This can be estimated by judgement or, more formally, by applying FMEA at the component level to decide whether each failure would be revealed by the diagnostics. Appendix 10 shows the detailed scoring criteria.

An exponential model is proposed to reflect the increasing difficulty in further reducing BETA as the score increases (as discussed in paragraph 8.2.3.f). This is reflected in the following equation:

$$\beta = 0.3 \exp(-3.4S/2624)$$

Because of the nature of this model, additional features (as perceived by any user) can be proposed in each of the categories. The model can then be modified. If subsequent field data indicate a change of relative importance between the categories then adjust the scores in each category so that the category totals reflect the new proportions, also ensuring that the total possible raw score ($S = 2624$) remains unaltered.

The model can best be used iteratively to test the effect of design, operating and maintenance proposals where these would alter the scoring. A BETA value can be assessed for a proposed equipment. Proposed changes can be reflected by altering the scores and recalculating BETA. The increased design or maintenance cost can be reviewed against the costs and/or savings in unavailability by re-running the RAMS predictions using the improved BETA. As with all RAMS predictions the proportional comparison of values rather than the absolute value is of primary value.

8.3 FAULT TREE ANALYSIS

8.3.1 The Fault Tree

A Fault Tree is a graphical method of describing the combinations of events leading to a defined system failure. In fault tree terminology the system failure mode is known as the *top event*.

The fault tree involves essentially three logical possibilities and hence two main symbols. These involve *gates* such that the inputs below gates represent failures. Outputs (at the top) of gates represent a propagation of failure depending on the nature of the gate. The three types are:

> *The OR gate* whereby any input causes the output to occur;
> *The AND gate* whereby all inputs need to occur for the output to occur;
> *The voted gate*, similar to the *AND* gate, whereby two or more inputs are needed for the output to occur.

Figure 8.6 shows the symbols for the *AND* and *OR* gates and also draws attention to their equivalence to reliability block diagrams. The *AND* gate models the redundant case and is thus equivalent to the parallel block diagram. The *OR* gate models the series case whereby any failure causes the top event. An example of a voted gate is shown in Figure 8.10.

For simple trees the same equations given in Section 8.1 on reliability block diagrams can be used and the difference is merely in the graphical method of modelling. In probability terms the *AND* gate involves multiplying probabilities of failure and the *OR* gate the addition rules given in Chapter 7. Whereas block diagrams model paths of success, the fault tree models the paths of failure to the top event.

Figure 8.6

A fault tree is constructed as shown in Figure 8.7 in which two additional symbols can be seen. The rectangular box serves as a place for the description of the gate below it. Circles, always at the furthest point down any route, represent the basic events which serve as the enabling inputs to the tree.

Figure 8.7

8.3.2 Calculations

Having modelled the failure logic, for a system, as a fault tree the next step is to evaluate the frequency of the top event. As with block diagram analysis, this can be performed, for simple trees, using the formulae from Section 8.1. More complex trees will be considered later.

The example shown in Figure 8.7 could be evaluated as follows. Assume the following basic event data:

	Failure rate (PMH)	MDT (hours)
PSU	100	24
Standby	500	168
Motor	50	168
Detector	5	168
Panel	10	24
Pump	60	24

The failure rate 'outputs' of AND gates G2 and G3 can be obtained from the formula $\lambda_1 \times \lambda_2 \times (MDT_1 + MDT_2)$. Where an AND gate is actually a voted gate, as, for example, two out of three, then again the formulae from Section 8.1 can be used. The outputs of the OR gates G1 and GTOP can be obtained by adding the failure rates of the inputs. Figure 8.8 has the failure rate and MDT values shown.

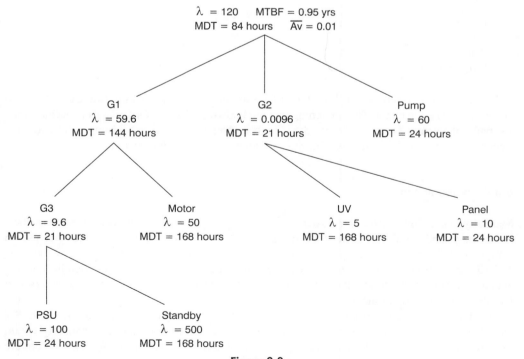

Figure 8.8

It often arises that the output of an OR gate serves as an input to another gate. In this case the MDT associated with the input would be needed for the equation. If the MDTs of the two inputs to the lower gate are not identical then it is necessary to compute an equivalent MDT. In Figure 8.8 this has been done for G1 even though the equivalent MDT is not needed elsewhere. It is the weighted average of the two MDTs weighted by failure rate. In this case,

$$\frac{(21 \times 9.6) + (168 \times 50)}{(9.6 + 50)} = 144 \text{ hours}$$

In the case of an AND gate it can be shown that the resulting MDT is obtained from the multiple of the individual MDTs divided by their sum. Thus for G3 the result becomes,

$$\frac{(24 \times 168)}{(24 + 168)} = 21 \text{ hours}$$

8.3.3 Cutsets

A problem arises, however, in evaluating more complex trees where the same basic initiating event occurs in more than one place. Using the above formulae, as has been done for Figure 8.8, would lead to inaccuracies because an event may occur in more than one *Cutset*. A Cutset is the name given to each of the combinations of base events which can cause the top event. In the example of Figure 8.7 the cutsets are:

    Pump
    Motor
    Panel and detector
    PSU and standby

The first two are referred to as First-Order Cutsets since they involve only single events which alone trigger the top event. The remaining two are known as Second-Order Cutsets because they are pairs of events. There are no third- or higher-order Cutsets in this example. The relative frequency of cutsets is of interest and this is addressed in the next section.

8.3.4 Computer tools

Manually evaluating complex trees, particularly with basic events which occur more than once, is not easy and would be time consuming. Fortunately, with the recent rapid increase in computer (PC) speed and memory capacity, a large number of software packages (such as TTREE) have become available for fault tree analysis. They are quite user-friendly and the degree of competition ensures that efforts continue to enhance the various packages, in terms of facilities and user-friendliness.

The majority of packages are sufficiently simple to use that even the example in Figure 8.7 would be considerably quicker by computer. The time taken to draw the tree manually would exceed that needed to input the necessary logic and failure rate data to the package. There are currently two methods of inputting the tree logic:

1. *Gate logic* which is best described by writing the gate logic for Figure 8.7 as follows:

GTOP   + G1 G2 PUMP
G1   + G3 MOTOR
G3   * PSU STANDBY
G2   * DETECT PANEL

+ represents an OR gate and * an AND gate. Each gate, declared on the right-hand side, subsequently appears on the left-hand side until all gates have been described in terms of all the basic events on the right. Modern packages are capable of identifying an illogical tree. Thus, gates which remain undescribed or are unreachable will cause the program to report an error.

2. *A graphical tree*, which is constructed on the PC screen by use of cursors or a mouse to pick up standard gate symbols and to assemble them into an appropriate tree.

Failure rate and mean down-time data are then requested for each of the basic events. The option exists to describe an event by a fixed probability as an alternative to stating a rate and down time. This enables fault trees to contain 'one-shot' events such as lightning and human error.

Most computer packages reduce the tree to Cutsets (known as minimal Cutsets) which are then quantified. Some packages compute by the simulation technique described in Section 9.5.

The outputs consist of:

GRAPHICS TO A PLOTTER OR PRINTER (e.g. Figures 8.7, 8.9, 8.10)
MTBF, AVAILABILITY, RATE (for the top event and for individual Cutsets)
RANKED CUTSETS
IMPORTANCE MEASURES

Cutset ranking involves listing the Cutsets in ranked order of one of the variables of interest – say, Failure Rate. In Figure 8.8 the Cutset whose failure rate contributes most to the top event is the PUMP (50%). The least contribution is from the combined failure of UV DETECTOR and PANEL. The ranking of Cutsets is thus:

PUMP   (50%)
MOTOR   (40%)
PSU and STANDBY   (8%)
UV DETECTOR and PANEL   (Negligible)

There are various applications of the *importance* concept but, in general, they involve ascribing a number either to the basic events or to Cutsets which describes the extent to which they contribute to the top event. In the example the PUMP MTBF is 1.9 years whereas the overall top event MTBF is 0.95 year. Its contribution to the overall failure rate is thus 50%. An importance measure of 50% is one way of describing the PUMP either as a basic event or, as is the case here, a Cutset.

If the cutsets were to be ranked in order of Unavailability the picture might be different, since the down times are not all the same. In Exercise 3, at the end of Chapter 9, the reader can compare the ranking by Unavailability with the above ranking by failure rate.

8.3.5 Allowing for CCF

Figure 8.9 shows the reliability block diagram of Figure 8.5 in fault tree form. The common cause failure can be seen to defeat the redundancy by introducing an OR gate above the redundant G1 gate.

Figure 8.9

Figure 8.10

Figure 8.10 shows another example, this time of 2 out of 3 redundancy, where a voted gate is used.

8.3.6 Fault tree analysis in design

Fault Tree Analysis, with fast computer evaluation (i.e. seconds/minutes, depending on the tree size), enables several design and maintenance alternatives to be evaluated. The effect of changes to redundancy or to maintenance intervals can be tested against the previous run. Again, evaluating the relative changes is of more value than obtaining the absolute MTBF.

Frequently a fault tree analysis identifies that the reduction of down time for a few component items has a significant effect on the top event MTBF. This can often be achieved by a reduction in the interval between preventive maintenance, which has the effect of reducing the repair time for dormant failures.

8.3.7 A Cautionary note

Problems can arise in interpreting the results of fault tress which contain only fixed probability events or a mixture of fixed probability and rate and time events.

If a tree combines fixed probabilities with rates and times then beware of the tree structure. If there are routes to the top of the tree (ie cutsets) which involve only fixed probabilities and, in addition, there are other routes involving rates and times then it is possible that the tree logic is flawed. This is illustrated by the example in Figure 8.11. G1 describes the scenario whereby

Figure 8.11

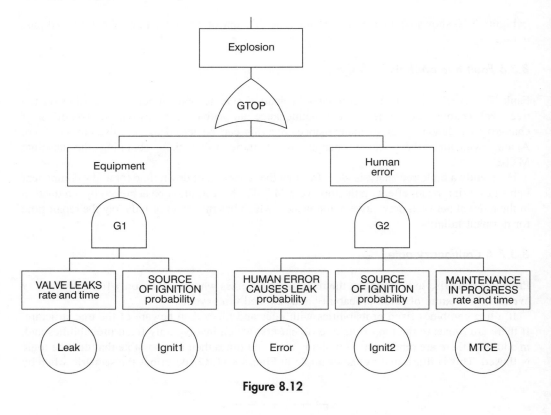

Figure 8.12

leakage, which has a rate of occurrence, meets a source of ignition. Its contribution to the top event is thus a rate at which explosion may occur. Conversely G2 describes the human error of incorrectly opening a valve and then meeting some other source of ignition. In this case, the contribution to the top event is purely a probability. It is in fact the probability of an explosion for each maintenance activity. It can be seen that the tree is not realistic and that a probability cannot be added to a rate. In this case a solution would be to add an additional event to G2 as shown in Figure 8.12. G2 now models the rate at which explosion occurs by virtue of including the maintenance activity as a rate (e.g. twice a year for 8 hours). G1 and G2 are now modelling failure in the same units (ie rate and time).

8.4 EVENT TREE DIAGRAMS

8.4.1 Why use Event Trees?

Whereas fault tree analysis (Section 8.3) is probably the most widely used technique for quantitative analysis, it is limited to AND/OR logical combinations of events which contribute to a single defined failure (the top event). Systems where the same component failures occurring in different sequences can result in different outcomes cannot so easily be modelled by fault trees. The fault tree approach is likely to be pessimistic since a fault tree acknowledges the occurrence of both combinations of the inputs to an AND gate whereas an Event Tree or Cause Consequence model can, if appropriate, permit only one sequence of the inputs.

8.4.2 The Event Tree model

Event Trees or Cause Consequence Diagrams (CCDs) resemble decision trees which show the likely train of events between an initiating event and any number of outcomes. The main element in a CCD is the decision box which contains a question/condition with YES/NO outcomes. The options are connected by paths, either to other decision boxes or to outcomes. Comment boxes can be added at any stage in order to enhance the clarity of the model.

Using a simple example the equivalence of fault tree and event tree analysis can be demonstrated. Figures 8.13 and 8.14 compare the fault tree AND and OR logic cases with their equivalent CCD diagrams. In both cases there is only one Cutset in the fault tree.

Figure 8.11(a) Pump 1 *and* 2
Figure 8.11(b) Smoke Detector *or* Alarm Bell

Figure 8.13

Figure 8.14

These correspond to the 'system fails' and 'no alarm' paths through the CCD diagrams in Figures 8.14(a) and (b), respectively.

Simple CCDs, with no feedback (explained later) can often be modelled using equivalent fault trees but in cases of sequential operation the CCD may be easier to perceive.

8.4.3 Quantification

A simple event tree, with no feedback loops, can be evaluated by simple multiplication of YES/NO probabilities where combined activities are modelled through the various paths.

Figure 8.15 shows the Fire Water Deluge example using a pump failure rate of 50 per million hours with a mean down time of 50 hours. The unavailability of each pump is thus obtained from:

$$50 \times 10^{-6} \times 50 = 0.0025$$

The probability of a pump not being available on demand is thus 0.0025 and the probabilities of both 100% system failure and 50% capacity on demand are calculated.

The system fail route involves the square of 0.0025. The 50% capacity route involves two ingredients of 0.0025 × 0.9975. The satisfactory outcome is, therefore, the square of 0.9975.

8.4.4 Differences

The main difference between the two models (fault tree and event tree) is that the event tree models the order in which the elements fail. For systems involving sequential operation it may well be easier to model the failure possibilities by event tree rather than to attempt a fault tree.

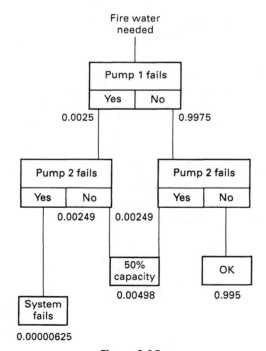

Figure 8.15

In the above example the event tree actually evaluated two possible outcomes instead of the single outcome (no deluge water) in the corresponding fault tree. As was seen in the example, the probabilities of each outcome were required and were derived from the failure rate and down time of each event.

The following table summarizes the main differences between Event Tree and Fault Tree models.

Cause Consequence	*Fault Tree*
Easier to follow for non-specialist	Less obvious logic
Permits several outcomes	Permits one top event
Permits sequential events	Static logic (implies sequence is irrelevant)
Permits intuitive exploration of outcomes	Top-down model requires inference
Permits feedback (e.g. waiting states)	No feedback
Fixed probabilities	Fixed probabilities and rates and times

8.4.5 Feedback loops

There is a complication which renders event trees difficult to evaluate manually. In the examples quoted so far, the exit decisions from each box have not been permitted to revisit existing boxes. In that case the simple multiplication of probabilities is not adequate.

Feedback loops are required for continuous processes or where a waiting process applies such that an outcome is reached only when some set of circumstances arises. Figure 8.16 shows a case where a feedback loop is needed where it is necessary to model the situation that a flammable liquid may or may not ignite before the relief valve closes. Either numerical integration or simulation (Section 9.5) is needed to quantify this model and a PC computer solution is preferred.

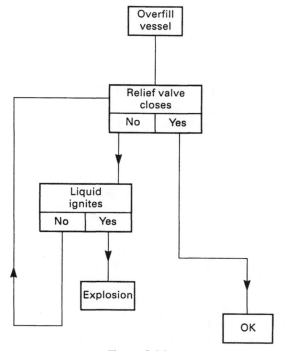

Figure 8.16

9 Quantifying the reliability models

9.1 THE RELIABILITY PREDICTION METHOD

This section summarizes how the methods described in Chapters 8 and 9 are brought together to quantify RAMS and Figure 9.1 gives an overall picture of the prediction process. It has already been emphasized that each specific system failure mode has to be addressed separately and thus targets are required for each mode (or top event in fault tree terminology). Prediction requires the choice of suitable failure rate data and this has been dealt with in detail in Chapter 4. Down times also need to be assessed and it will be shown, in Section 9.2, how repair times and diagnostic intervals both contribute to the down time. The probability of human error (Section 9.4) may also need to be assessed where fault trees or event trees contain human events (e.g. operator fails to act following an alarm).

One or more of the modelling techniques described in Chapter 8 will have been chosen for the scenario. The choice between block diagram and fault tree modelling is very much a matter for the analyst and will depend upon:

- which technique he/she favours
- what tools are available (i.e. FTA program)
- the complexity of the system to be modelled
- the most appropriate graphical representation of the failure logic

Chapter 8 showed how to evaluate the model in terms of random coincident failures. Common cause failures then need to be assessed as were shown in Section 8.2. These can be added into the models either as:

- series elements in the reliability block diagrams (Section 8.1)
- OR gates in the fault trees (Section 8.3.5)

Traditionally this process will provide a single predicted RAMS figure. However, the work described in Section 4.4 allows the possibility of expressing the prediction as a confidence range and showed how to establish the confidence range for mixed data sources. Section 9.6 shows how comparisons with the original targets might be made.

Figure 9.1 also reminds us that the opportunity exists to revise the targets should they be found to be unrealistic. It also emphasizes that the credibility of the whole process is dependent on field data being collected to update the data sources being used. The following sections address specific items which need to be quantified.

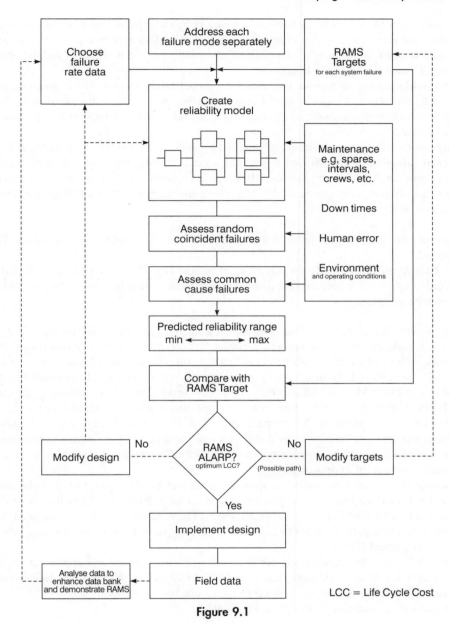

Figure 9.1

LCC = Life Cycle Cost

9.2 ALLOWING FOR DIAGNOSTIC INTERVALS

We saw, in Section 8.1.4, how the down time of unrevealed failures could be assessed. Essentially it is obtained from a fraction of the proof-test interval (i.e. half, at the unit level) as well as the MTTR (mean time to repair).

Some data bases include information about MTTRs and those that do have been indicated in Section 4.2.

In many cases there is both auto-test, whereby a programmable element in the system carries out diagnostic checks to discover unrevealed failures, as well as a manual proof-test. In practice the auto-test will take place at some relatively short interval (e.g. 8 minutes) and the proof-test at a longer interval (e.g. 4000 hours).

The question arises as to how the reliability model takes account of the fact that failures revealed by the auto-test enjoy a shorter down time than those left for the proof-test. The ratio of one to the other is a measure of the diagnostic coverage and is expressed as a percentage of failures revealed by the test.

Diagnostic coverage targets frequently quoted are 60%, 90% and 99% (e.g. in the IEC 61508 functional safety standard). At first this might seem a realistic range of diagnostic capability ranging from simple to comprehensive. However, it is worth considering how the existence of each of these coverages might be established. There are two ways in which diagnostic coverage can be assessed:

1. By test: In other words failures are simulated and the number of diagnosed failures counted.
2. By FMEA: In other words the circuit is examined (by FMEA described in Section 9.3) ascertaining, for each potential component failure mode, whether it would be revealed by the diagnostic program.

Clearly a 60% diagnostic could be demonstrated fairly easily by either method. Test would require a sample of only a few failures to reveal 60% or alternatively a 'broad brush' FMEA, addressing blocks of circuitry rather than individual components, would establish in an hour or two if 60% is achieved. Turning to 90% coverage, the test sample would now need to exceed 20 failures and the FMEA would require a component level approach. In both cases the cost and time begin to become onerous. For 99% coverage the sample size would now exceed 200 failures and this is likely to be impracticable. The alternative FMEA approach would be extremely onerous and involve several man-days and would be by no means conclusive.

The foregoing should be considered carefully before accepting the credibility of a target diagnostic coverage in excess of 90%.

Consider now a dual redundant configuration subject to 90% auto-test. Let the auto-test interval be 4 hours and the manual proof-test interval be 4380 hours. We will assume that the manual test reveals 100% of the remaining failures. The reliability block diagram needs to split the model into two parts in order to calculate separately in respect of the auto-diagnosed and manually diagnosed failures.

Figure 9.2 shows the parallel and common cause elements twice and applies the equations from Section 8.1 to each element. The total failure rate of the item, for the failure mode in question, is λ and the failure rates of the elements are shown below.

The equivalent fault tree is shown in Figure 9.3

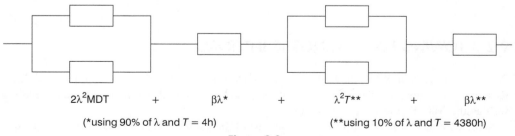

$2\lambda^2\mathrm{MDT}$ + $\beta\lambda*$ + $\lambda^2 T**$ + $\beta\lambda**$

(*using 90% of λ and $T = 4$h)   (**using 10% of λ and $T = 4380$h)

Figure 9.2

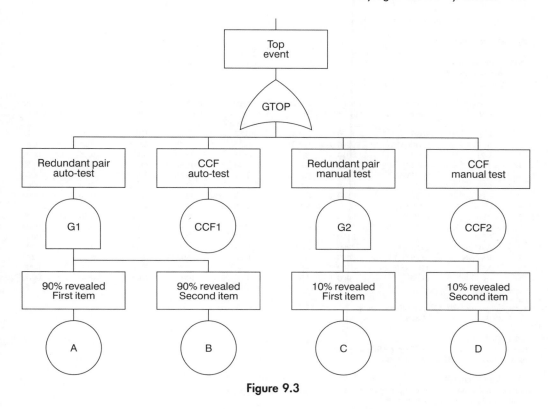

Figure 9.3

9.3 FMEA (FAILURE MODE AND EFFECT ANALYSIS)

The Fault Trees, Block diagrams and Event Tree models, described earlier, will require failure rates for the individual blocks and enabling events. This involves studying a circuit or mechanical assembly to decide how its component parts contribute to the overall failure mode in question.

This process is known as FMEA and consists of assessing the effect of each component part failing in every possible mode. The process consists of defining the overall failure modes (there will usually be more than one) and then listing each component failure mode which contributes to it. Failure rates are then ascribed to each component level failure mode and the totals for each of the overall modes are obtained.

The process of writing each component and its failures into rows and columns is tedious but PC programs are now available to simplify the process. Figure 9.4 is a sample output from the FARADIP.THREE package. Each component is entered by answering the interrogation for Reference, Name, Failure Rate, Modes and Mode Percentages. The table, which can be imported into most word-processing packages, is then printed with failure rate totals for each mode.

The concept of FMEA can be seen in Figure 9.4 by looking at the column headings 'Failure Mode 1' and 'Failure Mode 2'. Specific component modes have been assessed as those giving rise to the two overall modes (Spurious Output and Failure of Output) for the circuit being analysed.

Note that the total of the two overall failure mode rates is less than the parts count total. This is because the parts count total is the total failure rate of all the components for all of their failure

FARADIP-THREE PRINTOUT   18/12/96

DETECTOR CIRCUIT

FMEA Filename is : CCT22
Environment factor is : 1
Quality factor is : 1

	Comp't Ref	Comp't name	Total failure rate	Failure Mode 1	Mode 1 factor	Failure rate mode 1	Failure mode 2	Mode 2 factor	Failure rate mode 2
1	IC1	8086	.1500	LOW	0.80	.1200	HIGH	0.01	.0015
2	IC12	CMOSOA	.0800	HIGH	0.25	.0200	LOW	0.25	.0200
3	D21	SiLP	.0010	O/C	0.25	.0003	S/C	0.15	.0002
4	TR30	NPNLP	.0500	S/C	0.30	.0150	O/C	0.30	.0150
5	Z3	QCRYST	.1000	All	1.00	.1000	None	0.00	.0000
6	C9	TANT	.0050	S/C	1.00	.0050	None	0.00	.0000
7	*25	RFILM	.0500	O/C	0.50	.0250	None	0.00	.0000
8	UV3	UVDET	5.000	SPUR	0.50	2.500	FAIL	0.50	2.500
9	*150	CONNS	.0600	50%	0.50	.0300	50%	0.50	.0300
10	SW2	MSWTCH	.5000	O/C	0.30	.1500	S/C	0.10	.0500
11	PCB	PTPCB	.0100	20%	0.20	.0020	20%	0.20	.0020
12	R5COIL	COIL	.2000	COILOC	0.10	.0200	None	0.00	.0000
13	R5CONT	CONTCT	.2000	O/C	0.80	.1600	S/C	0.10	.0200
14	X1	TRANSF	.0300	All	1.00	.0300	None	0.00	.0000
15	F1	FUSE	.1000	All	1.00	.1000	None	0.00	.0000

Parts count
Total Failure rate = 6.536 per Million hours
Total MTBF   = 17.47 Years

SPURIOUS OUTPUT
Failure mode 1 rate   = 3.277 per Million Hours
Failure mode 1 MTBF = 34.83 Years

FAILURE OF OUTPUT
Failure mode 2 rate   = 2.639 per Million hours
Failure mode 2 MTBF = 43.26 Years

Figure 9.4

modes, whereas the specific modes being analysed do not cover all failures. In other words, there are component failure modes which do not cause either of the overall modes being considered.

Another package, FailMode from ITEM, enables FMEAs to be carried out to US Military Standard 1629A. This standard provides detailed step-by-step guidelines for performing FMEAs, including Criticality Analysis, which involves assigning a criticality rating to each failure.

The FMEA process does not enable one to take account of any redundancy within the assembly which is being analysed. In practice, this is not usually a problem, since small elements of redundancy can often be ignored, their contribution to the series elements being negligible.

9.4 HUMAN FACTORS

9.4.1 Background

It can be argued that the majority of well-known major incidents, such as Three Mile Island, Bhopal, Chernobyl, Zeebrugge and Clapham, are related to the interaction of complex systems with human beings. In short, the implication is that human error was involved, to a larger or greater extent, in these and similar incidents. For some years there has been an interest in modelling these factors so that quantified reliability and risk assessments can take account of the contribution of human error to the system failure.

As with other forms of reliability and risk assessment, the first requirement is for failure rate/ probability data to use in the fault tree or whichever other model is used. Thus, human error rates for various forms of activity are needed. In the early 1960s there were attempts to develop a database of human error rates and these led to models of human error whereby rates could be estimated by assessing relevant factors such as stress, training, complexity and the like. These human error probabilities include not only simple failure to carry out a given task but diagnostic tasks where errors in reasoning, as well as action, are involved. There is not a great deal of data available (see Section 9.4.6) since:

- Low probabilities require large amounts of experience in order for a meaningful statistic to emerge.
- Data collection concentrates on recording the event rather than analysing the causes.
- Many large organizations have not been prepared to commit the necessary resources to collect data.

More recently interest has developed in exploring the underlying reasons, as well as probabilities, of human error. In this way, assessments can involve not only quantification of the hazardous event but also an assessment of the changes needed to bring about a reduction in error.

9.4.2 Models

There are currently several models, each developed by separate groups of analysts working in this field. Whenever several models are available for quantifying an event the need arises to compare them and to decide which is the most suitable for the task in hand. Factors for comparison could be:

- *Accuracy* – There are difficulties in the lack of suitable data for comparison and validation.
- *Consistency* – Between different analysts studying the same scenario.
- *Usefulness* – In identifying factors to change in order to reduce the human error rate.
- *Resources* – Needed to implement the study.

One such comparison was conducted by a subgroup of the Human Factors in Reliability Group, and their report *Human Reliability Assessor's Guide* (SRDA R11), which addresses eight of the better-known models, is available from SRD, AEA Technology, Thomson House, Risley, Cheshire, UK WA3 6AT. The report is dated June 1995.

The following description of three of the available models will provide some understanding of the approach. A full application of each technique, however, would require a more detailed study.

9.4.3 HEART (Human Error Assessment and Reduction Technique)

This is a deterministic and fairly straightforward method developed by J. C. Williams during the early 1980s. It involves choosing a human error probability from a table of error rates and then modifying it by multiplication factors identified from a table of error-producing conditions. It is considered to be of particular use during design since it identifies error producing conditions and therefore encourages improvements. It is a quick and flexible technique requiring few resources. The error rate table, similar to that given in Appendix 6, contains nine basic error task types. It is:

Task	Probability of error
Totally unfamiliar, perform at speed, no idea of outcome	0.55
Restore system to new or original state on a single attempt without supervision or procedures checks	0.26
Complex task requiring high level of comprehension and skill	0.16
Fairly simple task performed rapidly or given scant attention	0.09
Routine highly practised, rapid task involving relatively low level of skill	0.02
Restore system to new state following procedure checks	0.003
Totally familiar task, performed several times per hour, well motivated, highly trained staff, time to correct errors	0.0004
Respond correctly when there is augmented supervisory system providing interpretation	0.00002
Miscellaneous task – no description available	0.03

The procedure then describes 38 'error-producing conditions' to each of which a maximum multiplier is ascribed. Any number of these can be chosen and, in turn, multiplied by a number between 0 and 1 in order to take account of the analyst's assessment of what proportion of the maximum to use. The modified multipliers are then used to modify the above probability. Examples are:

Error-producing condition	Maximum multiplier
Unfamilar with infrequent and important situation	×17
Shortage of time for error detection	×11
No obvious means of reversing an unintended action	×8
Need to learn an opposing philosophy	×6
Mismatch between real and perceived task	×4
Newly qualified operator	×3
Little or no independent checks	×3
Incentive to use more dangerous procedures	×2
Unreliable instrumentation	×1.6
Emotional stress	×1.3
Low morale	×1.2
Inconsistent displays and procedures	×1.2
Disruption of sleep cycles	×1.1

The following example illustrates the way the tables are used to calculate a human error probability.

Assume that an inexperienced operator is required to restore a plant bypass, using strict procedures but which are different to his normal practice. Assume that he is not well aware of the hazards, late in the shift and that there is an atmosphere of unease due to worries about impending plant closure.

The probability of error, chosen from the first table, might appropriately be 0.003.

5 error producing conditions might be chosen from the second table as can be seen in the following table.

For each condition the analyst assigns a 'proportion of the effect' from judgement (in the range 0-1).

The table is then drawn up using the calculation:

$$[(EPC - 1) \times (Proportion)] + 1$$

The final human error probability is the multiple of the calculated values in the table times the original 0.003.

FACTOR	EPC	PROPORTION EFFECT	$[(EPC{-}1) \times (Proportion)] + 1$
Inexperience	3	0.4	$[(3{-}1) \times (0.4)] + 1 = 1.8$
Opposite technique	6	1	6
Low awareness of risk	4	0.8	3.4
Conflicting objectives	2.5	0.8	2.2
Low morale	1.2	0.6	1.12

Hence ERROR RATE $0.003 \times 1.8 \times 6 \times 3.4 \times 2.2 \times 1.12 = \underline{0.27}$

Similar calculations can be performed at percentile bounds. The full table provides 5th and 95th percentile bands for the error rate table.

Note that since the probability of failure cannot exceed 1 and, therefore, for calculations taking the prediction above 1 it will be assumed that the error WILL almost certainly occur.

9.4.4 THERP (Technique for Human Error Rate Prediction)

This was developed by A. D. Swain and H. E. Guttmann and is widely used. The full procedure covers the definition of system failures of interest, through error rate estimation, to recommending changes to improve the system. The analyst needs to break each task into steps and then identify each error that can occur. Errors are divisible into types as follows:

Omission of a step or an entire task
Selects a wrong command or control
Incorrectly positions a control
Wrong sequence of actions
Incorrect timing (early/late)
Incorrect quantity

The sequence of steps is represented in a tree so that error probabilities can be multiplied along the paths for a particular outcome.

Once again (as with HEART), there is a table of error probabilities from which basic error rates for tasks are obtained. These are then modified by 'shaping parameters' which take account of stress, experience and other factors known to affect the error rates.

The analysis takes account of dependence of a step upon other steps. In other words, the failure of a particular action (step) may alter the error probability of a succeeding step.

9.4.5 TESEO (Empirical Technique To Estimate Operator Errors)

This was developed by G. C. Bellow and V. Colombari from an analysis of available literature sources in 1980. It is applied to the plant control operator situation and involves an easily applied model whereby live factors are identified for each task and the error probability is obtained by multiplying together the five factors as follows:

Activity
– Simple	0.001
Requires attention	0.01
Non-routine	0.1

Time stress (in seconds available)
– 2 (routine), 3 (non-routine)	10
10 (routine), 30 (non-routine)	1
20 (routine)	0.5
45 (non-routine)	0.3
60 (non-routine)	0.1

Operator
– Expert	0.5
Average	1
Poorly trained	3

Anxiety
– Emergency	3
Potential emergency	2
Normal	1

Ergonomic (i.e. *Plant interface*)
– Excellent	0.7
Good	1
Average	3–7
Very poor	10

Other methods
There are many other methods such as:

SLIM (Success Likelihood Index Method)
APJ (Absolute Probability Judgement)
Paired Comparisons
IDA (The Influence Diagram Approach)
HCR (Human Cognitive Reliability Correlation)

These are well described in the HFRG document mentioned above.

9.4.6 Human error rates

Frequently there are not sufficient resources to use the modelling approach described above. In those cases a simple error rate per task is needed. Appendix 6 is a table of such error rates which has been put together as a result of comparing a number of published tables.

One approach, when using such error rates in a fault tree or other quantified method, is to select a pessimistic value (the circumstances might suggest 0.01) for the task error rate. If, in the overall incident probability computed by the fault tree, the contribution from that human event is negligible then the problem can be considered unimportant. If, however, the event dominates the overall system failure rate then it would be wise to rerun the fault tree (or simulation) using an error rate an order less pessimistic (e.g. 0.001). If the event still dominates the analysis then there is a clear need for remedial action by means of an operational or design change. If the event no longer dominates at the lower level of error probability then there is a grey area which will require judgement to be applied according to the circumstances. In any case, a more detailed analysis is suggested.

A factor which should be kept in mind when choosing error rates is that human errors are not independent. Error rates are likely to increase as a result of previous errors. For instance, an audible alarm is more likely to be ignored if a previous warning gauge has been recently misread.

In the 1980s it was recognised that a human error data base would be desirable. In the USA the NUCLARR data base (see also Section 4.2.2.5) was developed and this consists of about 50% human error data although this is heavily dependent on expert judgement rather than solid empirical data. In the UK, there is the CORE-DATA (Computerized Operator Reliability and Error Database) which is currently being developed at the University of Birmingham.

9.4.7 Trends

Traditionally, the tendency has been to add additional levels of protection rather than address the underlying causes of error. More recently there is a focus of interest in analysing the underlying causes of human error and seeking appropriate procedures and defences to minimize or eliminate them.

Regulatory bodies, such as the UK Health and Safety Executive, are taking a greater interest in this area and questions are frequently asked about the role of human error in the hazard assessments which are a necessary part of the submissions required from operators of major installations (see Chapter 21).

9.5 SIMULATION

9.5.1 The technique

Block Diagram, Fault Tree and Cause Consequence analyses were treated, in Chapters 7–9, as deterministic methods. In other words, given that the model is correct then for given data there is only one answer to a model. If two components are in series reliability (fault tree OR gate) then, if each has a failure rate of 5 per million hours, the overall failure rate is 10 per million hours – no more, no less. Another approach is to perform a computer-based simulation, sometimes known as Monte Carlo analysis, in which random numbers are used to sample from probability distributions.

In the above example, two random distributions each with a rate of 5 per million would be set up. Successive time slots would be modelled by sampling from the two distributions in order to ascertain if *either* distribution yielded a failure in that interval.

One approach, known as event-driven simulation inverts the distribution to represent time as a function of the probability of a failure occurring. The random number generator is used to provide a probability of failure which is used to calculate the time to the next failure. The events generated in this manner are then logged in a 'diary' and the system failure distribution derived from the component failure 'diary'. As an example assume we wish to simulate a simple exponential distribution then the probability of failing in time t is given by:

$$R(t) = e^{-\lambda t}$$

Inverting the expression we can say that:

$$t = (\log_e R)/\lambda$$

Since R is a number between 0 and 1 the random number generator can be used to provide this value which is divided by λ to provide the next value of t. The same approach is adopted for more complex expressions such as the Weibull.

A simulation would be run many thousands of times and the overall rate of system failure counted. This might be 10 per million of 9.99998 or 10.0012 and, in any case, will yield slightly different results for each trial. The longer each simulation run and the more runs attempted, the closer will the ultimate answer approximate to 10 per million.

This may seem a laborious method for assessing what can be obtained more easily from deterministic methods. Fault Tree, Cause Consequence and Simple Block diagram methods are, however, usually limited to simple AND/OR logic and *constant failure rates* and straightforward *mean down times*.

Frequently problems arise due to complicated failure and repair scenarios where the effect of failure and the redundancy depend upon demand profiles and the number of repair teams. Also, it may be required to take account of failure rates and down times which are not constant. The assessment may therefore involve:

- Log Normal down times
- Weibull down times
- Weibull (not constant) failure rates
- Standby items with probabilities of successful start
- Items with variable profiles for which the MTBF varies with some process throughput
- Spares requirements
- Maintenance skill types and quantities
- Logistical delays
- Ability to make-up lost availability within defined rules and limits

It is not easy to evaluate these using the techniques already explained in this chapter and, for them, simulation now provides a quick and cost-effective method.

One drawback to the technique is that the lower the probability of the events, the greater the number of trials that are necessary in order to obtain a satisfactory result. The other, which has recently become much reduced, was the program cost and computer run times involved, which were greatly in excess of fault tree and block diagram approaches. With ever-increasing PC power there are now a number of cost-effective packages which can rival the deterministic techniques.

A recent development in reliability simulation (see Section 9.5.2) is the use of genetic algorithms. This technique enables modelling options (e.g. additional redundant streams) to be specified in the simulation. The algorithm then develops and tests the combinations of possibilities depending on the relative success of the outcomes.

There are a variety of algorithms for carrying out this approach and they are used in the various PC packages which are available. Some specific packages are described in the following sections. There are many similarities between them.

9.5.2. Some packages

OPTAGON

This package was developed by British Gas Research and Development at their GRTC (Gas Research Technology Centre), Loughborough. It is a development of the earlier packages SWIFT and PARAGON and is primarily intended for modelling parallel throughputs where the variables listed in the bullet points in Section 9.5.1 above are of concern.

After each simulation reported data, with means and standard deviations, include:

- System failure rate expressed as the number of periods of shortfall over the simulation.
- Shortfall as a proportion of the total demand.
- Unavailability, being the proportion of the simulation time when a shortfall is present.
- Total cost of shortfall, capital, operating, maintenance and spares costs which can be adjusted by a discount factor over time.

A particular feature is the use of the Genetic Algorithms already mentioned. These apply the Darwinian principle of natural selection by searching for the optimal solution emerging from the successive simulation runs. It is achieved by expressing the characteristics of a system (such as the bullet points listed above) in the form of a binary string. The string (known as a gene-string) can then be created by random number generation. A weighting process is used to favour those genes which lead to the more optimistic outcomes by increasing the probability of their choice in successive simulations.

MAROS and TARO

MAROS is an early simulation tool from Jardine, being a RAM simulation tool with features which include networking, in-line buffering, complex production operations, sales contract shortfalls, batch export (shipping), cause and effect logic (dynamic fault-tree). TARO is the follow-on generation of asset management simulation tools (Total Asset Review and Optimization) which focus on the operations phase of a project to manage maintenance and operations while addressing system performance requirements. Currently there are three industry applications:

1. **Oil & Gas:** Enhanced MAROS dealing with multiple-products, more complex and detailed maintenance and logistics.
2. **Petro-chemical & Refining:** A product for petro-chemical and refinery applications, dealing with complex multiproduct production operations in conjunction with reliability and maintenance issues. It handles product blending, disposal management, complex networked buffering, turnaround planning. Provides the ability to optimise unit and tank capacities in-line with feedstock purchases, slate definitions and product revenue predictions.
3. **Railways & Integrated Transport:** Performance simulation for railways and other transport. Simulates a prescribed timetable of complex traffic flow, predicting punctuality

and identifying reasons for delays, as a function of infrastructure reliability and maintainability. Includes life-cycle costing and asset management features common to all TARO products.

All the tools apply direct simulation (next-event) techniques using proprietary algorithms specifically designed for speed of execution, thus enabling modelling of large complex systems. The contact is Hari.Vamadevan@jardine.co.uk.

RAM4

This Monte Carlo package, available from Fluor Global Services, allows the user to construct reliability block diagrams and to import them, if desired, from spread sheets. It provides outputs in histogram form for importing into reports as, for example, a distribution of times to repair for a given simulation.

An element data base can be set up for rapid entry of specific items with stated repair times, spares types, failure distributions etc. Maintenance types can be specified allowing failed items to be put in a queue awaiting the appropriate skill. Fluor Global are at Farnham, Surrey (www.FluorGS.co.uk).

ITEM Toolkit

This is also a Monte Carlo package based on reliability block diagrams. It copes with revealed and unrevealed failures, preventive and corrective maintenance regimes, ageing and maintenance queuing. The usual standby and start-up scenarios are modelled and non-random distributions for failure rate and down time can be modelled. System performance is simulated over a number of life-cycles to predict unavailability, number of system failures and required spares levels. ITEM are at Fareham, Hants (www.itemuk.com)

9.6 COMPARING PREDICTIONS WITH TARGETS

In the light of the work described in Section 4.4 we saw that is now possible to attempt some correlation between predicted and field reliability and that the confidence in the prediction depends upon the data used.

The studies referred to indicated that the results were equally likely to be optimistic as pessimistic. Therefore one interpretation is that we are 50% confident that the field result will be equal to or better than the predicted RAMS value. However, a higher degree of confidence may be desired, particularly if the RAMS prediction is for a safety-related failure mode. If industry specific data has been used for the prediction and 90% confidence is required then, consulting the tables in Section 4.4, a failure rate of 4 times the predicted value would be used.

If the '4 times failure rate' figure is higher than the value which coincides with the Maximum Tolerable Risk, discussed in Sections 3.3 and 10.2, then the proposed design is not acceptable. If it falls between the Maximum Tolerable and Broadly Acceptable Risk values then the ALARP principle is applied as shown in the cost per life saved examples in Section 3.3.

The following is an example:

The maximum tolerable risk of fatality associated with a particular system failure mode might be 10^{-4} per annum. The failure rate, for that mode, which risk assessment shows is associated with that frequency, is say 10^{-3} failures per annum. If the broadly acceptable risk is 10^{-6} per annum then it follows that it will be achieved with a failure rate 100 times less, 10^{-4} per annum.

Let the predicted failure rate (using industry specific data) for the system failure mode in question be 2×10^{-4} per annum and assume that we wish to be 90% confident in the result. As discussed above the 90% confidence failure rate would be $4 \times 2 \times 10^{-4} = 8 \times 10^{-4}$ per annum. In other words we can be 90% sure that the field failure rate will be better than 8×10^{-4} per annum (in other words a fatality risk of 8×10^{-5} per annum).

This is better than the maximum tolerable risk but not small enough to be 'dismissed' as broadly acceptable. Therefore, a design proposal is made (perhaps additional redundancy at a cost of £5000) to improve the failure rate. Assume that the outcome is a new predicted failure rate of 10^{-4} per annum (i.e. 4×10^{-4} at 90% confidence which is 4×10^{-5} per annum risk of fatality).

Assuming 2 fatalities and a 40 year system life, the cost per life saved calculation is:
£5000/([$8 \times 10^{-5} - 4 \times 10^{-5}$] $\times$ 2 $\times$ 40) = £1.5 million

If this exceeds the cost per life saved criteria being applied (see Section 3.3.2) then the existing design would be considered to offer an integrity which is ALARP. If not then the design proposal would need to be considered.

EXERCISES

1. The reliability of the two-valve example of Figure 2.1 was calculated, for two failure modes, in Sections 7.3 and 7.4. Imagine that, to improve the security of supply, a twin parallel stream is added as follows:

   Construct reliability block diagrams for:

   (a) Loss of supply
   (b) Failure to control downstream over-pressure

   and recalculate the reliabilities for one year.

2. For this twin-stream case, imagine that the system is inspected every two weeks, for valves which have failed shut. How does this affect the system failure rate in respect of loss of supply?

3. In Section 8.3, the Cutsets were ranked by failure rate. Repeat this ranking by unavailability.

10 Risk assessment (QRA)

10.1 FREQUENCY AND CONSEQUENCE

Having identified a hazard, the term 'risk analysis' is often used to embrace two assessments:

- The frequency (or probability) of the event
- The consequences of the event

Thus, for a process plant the assessments could be:

- The probability of an accidental release of a given quantity of toxic (or flammable) material might be 1 in 10 000 years.
- The consequence, following a study of the toxic (or thermal radiation) effects and having regard to the population density, might be 40 fatalities.

Clearly these figures describe an unacceptable risk scenario, irrespective of the number of fatalities.

The term QRA (Quantified Risk Assessment) refers to the process of assessing the frequency of an event and its measurable consequences (eg fatalities, damage).

The analysis of consequence is a specialist area within each industry and may be based on chemical, electrical, gas or nuclear technology. Prediction of frequency is essentially the same activity as reliability prediction, the methods for which have been described in Chapters 7 to 9. In many cases the method is identical, particularly where the event is dependent only on:

- Component failures
- Human error
- Software

Both aspects of quantitative risk assessment are receiving increased attention, particularly as a result of Lord Cullen's inquiry into the Piper Alpha disaster.

Risk analysis, however, often involves factors such as lightning, collision, weather factors, flood, etc. These are outlined in Section 10.4.

10.2 PERCEPTION OF RISK AND ALARP

The question arises of setting a quantified level for the risk of fatality. The meaning of such words as 'tolerable', 'acceptable' and 'unacceptable' becomes important. There is, of course, no such thing as zero risk and it becomes necessary to think about what levels are 'tolerable' or even 'acceptable'.

In this context acceptable is generally taken to mean that we accept the probability of fatality as reasonable, having regard to the circumstances, and would not seek to expend much effort in reducing it further.

Tolerable, on the other hand, implies that whilst we are prepared to live with the particular risk level we would continue to review its causes and the defences we might take with a view to reducing it further. Cost would probably come into the picture in that any potential reduction in risk would be compared with the cost needed to achieve it.

Unacceptable means that we would not tolerate that level of risk and would not participate in the activity in question nor permit others to operate a process that exhibited it.

The principle of ALARP (As low as reasonably practicable) describes the way in which risk is treated legally and by the HSE (Health and Safety Executive) in the UK. The concept is that all reasonable measures will be taken in respect of risks which lie in the 'tolerable' zone to reduce them further until the cost of further risk reduction is grossly disproportionate to the benefit.

It is at this point that the concept of 'cost per life saved' (described in Chapter 3) arises. Industries and organizations are reluctant to state specific levels of 'cost per life saved' which they would regard as becoming grossly disproportionate to a reduction in risk. Nevertheless figures in the range £500 000 to £2 000 000 are not infrequently quoted. The author has seen £750 000 for government evaluation of road safety proposals and £2 000 000 for rail safety quoted at various times. It is also becoming the practice to increase the Cost per Life Saved figure in the case of multiple fatalities. Thus, for example, if £1 000 000 per Life Saved is normally used for a single fatality then £20 000 000 per Life Saved might well be applied to a 10 death scenario.

Perception of risk is certainly influenced by the circumstances. A far higher risk is tolerated from voluntary activities than involuntary risks (people feel that they are more in control of the situation on roads than on a railway). Compare, also, the fatality rates of cigarette smoking and those associated with train travel in Appendix 7. They are three orders of magnitude apart. Furthermore the risk associated with multiple, as opposed to single, fatalities is expected to be much lower in the former case.

It is sometimes perceived that the risk which is acceptable to the public is lower than that to an employee who is deemed to have chosen to work in a particular industry. Members of the public, however, endure risks without consent or even knowledge.

Another factor is the difference between individual and societal risk. An incident with multiple fatalities is perceived as less acceptable than the same number occurring individually. Thus, whereas the 10^{-4} level is tolerated for vehicles (single death – voluntary), even 10^{-6} might not satisfy everyone in the case of a nuclear power station incident (multiple deaths – involuntary).

Figure 10.1, shows how, for a particular industry or application the Intolerable, Tolerable and Acceptable regions might be defined and how they can be seen to reduce as the number of fatalities increases. Thus for a single fatality (left hand axis) risks of 10^{-5} to 10^{-3} are regarded as ALARP. Above 10^{-3} is unacceptable and below 10^{-5} is acceptable. For 10 fatalities,however, the levels are 10 times more stringent.

This topic is treated very fully in the HSE publication, *The Tolerability of Risk from Nuclear Power Stations*, upon which Figure 10.1 is based, and its more recent *Reducing Risks Protecting People*.

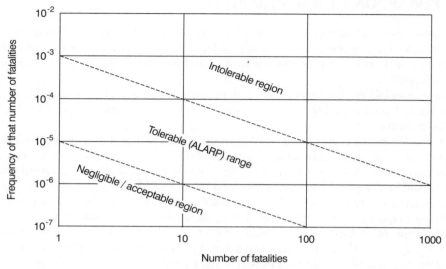

Figure 10.1 (Example only)

10.3 HAZARD IDENTIFICATION

Before an event (failure) can be quantified it must first be identified and there are a number of formal procedures for this process. HAZID (Hazard Identification) is used to identify the possible hazards, HAZOP (Hazard and Operability Study) is used to establish how the hazards might arise in a process whereas HAZAN (Hazard Analysis) refers to the process of analysing the outcome of a hazard. This is known as Consequence Analysis.

This is carried out at various levels of detail from the earliest stages of design throughout the project design cycle.

Preliminary Hazard Analysis, at the early system design phase, identifies safety critical areas, identifies and begins to quantify hazards and begins to set safety targets. It may include:

> Previous experience (historical information)
> Review of hazardous materials, energy sources, etc.
> Interfaces with operators, public, etc.
> Applicable legislation, standards and regulations
> Hazards from the environment
> Impact on the environment
> Software implications
> Safety-related equipment

More detailed Hazard Analysis follows in the detailed design stages. Now that specific hardware details are known and drawings exist, studies can address the effects of failure at component and software level. FMEA and Fault Tree techniques (Chapter 8) as well as HAZOP and Consequence Analyses are applicable here.

10.3.1 HAZOP

HAZOP (Hazard and Operability Studies) is a technique, developed in the 1970s, by loss prevention engineers working for Imperial Chemical Industries at Tees-Side UK. The purpose of a HAZOP is to identify hazards in the process. At one time this was done by individuals or groups of experts at a project meeting. This slightly blinkered approach tended to focus on the more obvious hazards and those which related to the specific expertise of the participants. In contrast to this, HAZOP involves a deliberately chosen balanced team using a systematic approach. The method is to systematically brainstorm the plant, part by part, and to review how deviations from the normal design quantities and performance parameters would affect the situation. Appropriate remedial action is then agreed.

One definition of HAZOP has been given as:

A Study carried out by a *Multidisciplinary Team*, who apply *Guidewords* to identify *Deviations* from the Design Intent of a system and its Procedures. The team attempt to identify the *Causes and Consequences* of these Deviations and the Protective Systems installed to minimize them and thus to make *Recommendations* which lead to Risk Reduction.

This requires a full description of the design (up-to-date engineering drawings, line diagrams, etc.) and a full working knowledge of the operating arrangements. A HAZOP is thus usually conducted by a team which includes designers and operators (including plant, process and instrumentation) as well as the safety (HAZOP) engineer.

A typical small process plant might be 'HAZOPed' by a team consisting of:

Chemical Engineer
Mechanical Engineer
Instrument Engineer
Loss Prevention (or Safety or Reliability) Engineer
Chemist
Production Engineer/Manager
Project Manager

A key feature is the HAZOP team leader who must have experience of HAZOP and be full time in the sense that he attends the whole study whereas some members may be part time. An essential requirement for the leader is experience of HAZOP in other industries so as to bring as wide a possible view to the probing process. Detailed recording of problems and actions is essential – during the meeting. Follow up and review of actions must also be formal. There must therefore be a full time Team Secretary who records all findings and actions.

The procedure will involve:

Define the scope and objectives of the HAZOP
Define the documentation required
Select the team
Prepare for the HAZOP (pre-reading)
Carry out and record the HAZOP
Implement the follow-up action
Record results

In order to formalize the analysis a 'guideword' methodology has evolved in order to point the analysts at the types of deviation. The guidewords are applied to each of the process parameters such as flow, temperature, pressure, etc. under normal operational as well as start-up and shut-down modes. Account should be taken of safety systems which are allowed, under specified circumstances, to be temporarily defeated. The following table describes the approach:

Guideword	Meaning	Explanation
NO or NOT	The parameter is zero	Something does not happen but no other effect
MORE THAN or LESS THAN	There are increases or decreases in the process parameter	Flows and temperatures are not normal
AS WELL AS	Qualitative Increase	Some additional effect
PART OF	Qualitative Increase	Partial effect (not all)
THE REVERSE	Opposite	Reverse flow or material
OTHER THAN	Substitution	Totally different effect

Each deviation of a parameter must have a credible *cause*, typically a component or human error related failure or a deviation elsewhere in the plant. Examples of typical *causes* might be:

DEVIATION	CAUSE
More Flow	Line rupture Control valve fail 'open'
Less Flow	Control valve fail 'closed' Leaking vessel or heat exchanger
No Flow	Blockage Rupture
Reverse Flow	Siphoning Check-valve failure
More Pressure	Restricted Flow Boiling
Less/No Pressure	Excessive Flow out Insufficient Flow in
More Level	Operator error Vessel leak
Less/No Level	Drain left open High barometric pressure

More Temperature	Loss of cooling
	Latent heat release
Less Temperature	Joule-Thomson cooling
	Adiabatic expansion
Part Composition	Loss of ratio control
	Dosing pump failure
More Composition	Carry-over
	By-products

Causes lead to *Consequences* which need to be assessed. When a parameter has varied beyond the design intent then it might lead to vessel rupture, fire, explosion toxic release, etc.

The *likelihood* may also be assessed. The reliability prediction techniques described earlier in this book, can be used to predict the frequency of specific events. However, these techniques may be reserved for the more severe hazards. In order to prioritize, a more qualitative approach at the HAZOP stage might be to assign, using team judgement only, say 5 grades of likelihood as for example;

1. Not more than once in the plant life
2. Up to once in 10 years
3. Up to once in 5 years
4. Up to once a year
5. More frequent than annually

A similar approach can be adopted for classifying *Severity* pending more formal quantification of the more severe consequences. The ranking might be:

1. No impact on plant or personnel
2. Damage to equipment only or minor releases
3. Injuries to unit personnel (contained on-site)
4. Major damage, limited off-site consequences
5. Major damage and extensive off-site consequences

One approach is to use a *Risk Matrix* to combine the *Likelihood* and *Severity* assessments in order to prioritize items for a more quantified approach and for further action. One such approach is:

	Severity 1	*Severity 2*	*Severity 3*	*Severity 4*	*Severity 5*
Likelihood 1	1	2	3	4	5
Likelihood 2	2	4	6	7	8
Likelihood 3	3	6	7	8	9
Likelihood 4	4	7	8	9	10
Likelihood 5	5	8	9	10	10

where '10' is the highest ranking of consequence and '1' is the lowest.

HAZOP was originally applied to finalized plant design drawings. However, changes arising at this stage can be costly and the technique has been modified for progressive stages of application throughout the design cycle. As well as being a design tool HAZOP can be equally successfully applied to existing plant and can lead to worthwhile modifications to the maintenance procedures.

Typical phases of the life-cycle at which HAZOP might be applied are:

> CONCEPTUAL DESIGN
> DETAILED DESIGN
> APPROVED FOR CONSTRUCTION
> 'AS-BUILT'
> PROPOSED MODIFICATIONS
> REGULATORY REQUIREMENTS

HAZOP can be applied to a wide number of types of equipment including:

> Process plant
> Transport systems
> Data and programmable systems (see UK DEF STD 0058)
> Buildings and structures
> Electricity generation and distribution
> Mechanical equipment
> Military equipment

In summary, HAZOP study not only reveals potential hazards but leads to a far deeper understanding of a plant and its operations.

Appendix 11 provides a somewhat simple example of a HAZOP.

10.3.2 HAZID

Whereas HAZOP is an open-ended approach, HAZID is a checklist technique. At an early stage, such as the feasibility study for a hazardous plant, HAZID enables the major hazards to be identified. At the conceptual stage a more detailed HAZID would involve designing out some of the major problems.

Often, the HAZID uses a questionnaire approach and each organization tends to develop and evolve its own list, based on experience. Appendix 12 gives an example of such a list and is reproduced by kind permission of the Institution of Gas Engineers (guidance document SR24).

10.3.3 HAZAN (Consequence Analysis)

This technique is applied to selected hazards following the HAZOP and HAZID activities. It is usually the high-consequence activities such as major spillage of flammable or toxic materials or explosion which are chosen. High-consequence scenarios usually tend to be the low-probability hazards.

Consequence analysis requires a detailed knowledge of the materials/hazards involved in order to predict the outcome of the various failures. The physics and chemistry of the outcomes is necessary in order to construct mathematical models necessary to calculate the effects on objects and human beings. Some examples are:

Flammable and toxic releases (heat radiation, food/water pollution and poisoning)
Structural collapse
Vehicle, ships and other impact (on structures and humans)
Nuclear contamination
Explosion (pressure vessels and chemicals)
Large scale water release (vessels, pipes and dams)

Reference to specific literature, in each case, is necessary.

10.4 FACTORS TO QUANTIFY

The main factors which may need to be quantified in order to assess the frequency of an event are as follows.

10.4.1 Reliability

Chapters 7 to 9 cover this element in detail.

10.4.2 Lightning and thunderstorms

It is important to differentiate between thunderstorm-related damage, which affects electrical equipment by virtue of induction or earth currents, and actual lightning strikes. The former is approximately one order (ten times) more frequent.

BS 6651: 1990 indicates an average of 10 thunderstorm days per annum in the UK. This varies, according to the geography and geology of the area, between 3 and 21 days per annum. Thunderstorm damage (usually electrical) will thus be related to this statistic. Some informal data suggest damage figures such as:

- Five incidents per square kilometre per annum where electrical equipment is used in outdoor or unprotected accommodation.
- 0.02 incidents per microwave tower.

Lightning strike, however, is a smaller probability and the rate per annum is derived by multiplying the effective area in square kilometres by the strikes per annum per square kilometre in Figure 10.2 (reproduced by kind permission of the British Standards Institution). The average is in the area of 0.3–0.5 per annum.

The effective area is obtained by subtending an angle of 45° around the building or object in question. Figure 10.3 illustrates the effect upon one elevation of a square building of side 10 m and height 2 m. The effective length is thus 14 m (10 + 2 + 2). BS 6651: 1990, from which Figure 10.2 is reproduced, contains a fuller method of assessment.

It must not be inferred, automatically, that a strike implies damage. This will depend upon the substance being struck, the degree of lightning protection and the nature of the equipment contained therein.

Figure 10.2 Number of lightning flashes to the ground per km² per year for the UK

Figure 10.3

10.4.3 Aircraft impact

Aircraft crash is a high-consequence but a low-probability event. The data are well recorded and a methodology has evolved for calculating the probability of impact from a crashing aircraft according to the location of the assessment in question. This type of study stemmed from concerns when carrying out risk assessments of nuclear plant but can be used for any other safety study where impact damage is relevant. Crashes are considered as coming from two causes.

Background

This is the 'ambient' source of crash, assumed to be randomly distributed across the UK. Likelihoods are generally taken as:

Type	Crash rate $\times 10^{-5}$ years per square mile
Private aircraft	7.5
Helicopters	2.5
Small transport aircraft	0.25
Airline transport	1.3
Combat military	3.8
TOTAL (all types)	15

Airfield proximity

These are considered as an additional source to the background and a model is required which takes account of the orientation from and distance to the runway. The probability of a crash, per square mile, is usually modelled as:

For take-off $\quad D = 0.22/r \times e^{-r/2} \times e^{-t/80}$
For landing $\quad D = 0.31/r \times e^{-r/2.5} \times e^{-t/43}$

where r is the distance in miles from the runway and t is the angle in degrees.

A full description of these techniques can be found in the CEGB publication GD/PE-N/403, which addresses the aircraft crash probabilities for Sizewell 'B'. A computer program for calculating crash rates (Prang) is available from SRD of AEA technology.

10.4.4 Earthquake

Earthquake intensities are defined according to Mercalli and the modified scale can be summarized as follows:

Intensity	Effect
I	Not felt.
II	Felt by persons at rest on upper floors.
III	Felt indoors. Hanging objects swing. Vibration similar to light trucks passing. May not be recognized as an earthquake.

IV   Hanging objects swing. Vibration like passing of heavy trucks or jolt sensation like heavy ball striking wall. Parked motor cars rock. Windows, dishes and doors rattle. Glasses and crockery clink. Wooden walls may creak.

V   Felt outdoors. Sleepers awakened. Liquids disturbed and some spilled. Small unstable objects displaced or upset. Pendulum clocks affected. Doors, pictures, etc. move.

VI   Felt by all. People frightened, run outdoors and walk unsteadily. Windows, dishes, glassware broken. Items and books off shelves. Pictures off walls and furniture moved or overturned. Weak plaster and masonry D cracked. Small bells ring, trees or bushes visibly shaken or heard to rustle.

VII   Difficult to stand. Noticed by drivers of motor cars. Hanging objects quiver. Furniture broken, damage to masonry D including cracks. Weak chimneys broken at roof line. Plaster, loose bricks, stones, tiles, etc. fall and some cracks to masonry C. Waves on ponds. Large bells ring.

VIII   Steering of motor cars affected. Damage or partial collapse of masonry C. Some damage to masonry B but not A. Fall of stucco and some masonry walls. Twisting and falling chimneys, factory stacks, elevated tanks and monuments. Frame houses moved on foundations if not secured. Branches broken from trees and cracks in wet ground.

IX   General panic. Masonry D destroyed and C heavily damaged (some collapse) and B seriously damaged. Reservoirs and underground pipes damaged. Ground noticeably cracked.

X   Most masonry and some bridges destroyed. Dams and dikes damaged. Landslides. Railway lines slightly bent.

XI   Rails bent. Underground pipelines destroyed.

XII   Total damage. Large rocks displaced. Objects in air.

The masonry types referred to are:

D   Weak materials, poor workmanship.
C   Ordinary materials and workmanship but not reinforced.
B   Good workmanship and mortar. Reinforced.
A   Good workmanship and mortar and laterally reinforced using steel, concrete, etc.
The range of interest is V to VIII since, below V the effect is unlikely to be of concern and above VIII the probability of that intensity in the UK is negligible.

The following table of frequencies is assumed to apply across the UK:

Intensity	Annual propability
V	12 $\times 10^{-3}$
VI	3.5 $\times 10^{-3}$
VII	0.7 $\times 10^{-3}$
VIII	0.075$\times 10^{-3}$

A useful reference is *Elementary Seismology*, by C. F. Richter (Freeman).

10.4.5 Meteorological factors

The Meteorological Office publishes a range of documents giving empirical data. by place and year, covering:

- Extreme wind speeds and directions
- Barometric pressure
- Snow depth
- Temperature
- Precipitation

These can be obtained from HMSO (Her Majesty's Stationery Office) and may be consulted in modelling the probability of extreme conditions which have been identified as being capable of causing the event in question.

10.4.6 Other Consequences

As a result of extensive measurements of real events, models have been developed to assess various consequences. The earlier sections have outlined specific examples such as lightning, earthquake and aircraft impact. Other events which are similarly covered in the appropriate literature and by a wide range of computer programs (see Appendix 10) are:

Chemical release
Gas explosion
Fire and blast
Ship collision
Pipeline corrosion
Pipeline rupture
Jet dispersion
Thermal radiation
Pipeline impact
Vapour cloud/pool dispersion

Part Four
Achieving Reliability and Maintainability

11 Design and assurance techniques

This chapter outlines the activities and techniques, in design and operation, which can be used to optimize reliability.

11.1 SPECIFYING AND ALLOCATING THE REQUIREMENT

The main objective of a reliability and maintainability programme is to assure adequate performance consistent with minimal maintenance costs. This can be achieved only if, in the first place, objectives are set and then described by suitable parameters. The intended use and environment of a system must be accurately stated in order to set realistic objectives and, in the case of contract design, the customer requirements must be clearly stated. It may well be that the customer has not considered these points and guidance may be necessary in persuading him or her to set reasonable targets with regard to the technology, environment and overall cost envisaged. Appropriate parameters have then to be chosen.

System reliability and maintainability will be specified, perhaps in terms of MTBF and MTTR, and values have then to be assigned to each separate unit. Thought must be given to the *allocation* of these values throughout the system such that the overall objective is achieved without over-specifying the requirement for one unit while under-specifying for another. Figure 11.1 shows a simple system comprising two units connected in such a way that neither may fail if the system is to perform. We saw in Chapter 7 that the system MTBF is given by:

$$\theta_s = \frac{\theta_1 \theta_2}{\theta_1 + \theta_2}$$

Figure 11.1

If the design objective for θ_s is 1000 h then this may be met by setting θ_1 and θ_2 both at 2000 h. An initial analysis of the two units, however, could reveal that unit 1 is twice as complex as, and hence likely to have half the MTBF of, unit 2. If the reliability is allocated equally, as suggested, then the design task will be comparatively easy for unit 2 and unreasonably difficult for unit 1. Ideally, the allocation of MTBF should be weighted so that:

$$2\theta_1 = \theta_2$$

Hence

$$\theta_s = \frac{2\theta_1^2}{3\theta_1} = \frac{2\theta_1}{3} = 1000 \text{ h}$$

Therefore

$$\theta_1 = 1500 \text{ h}$$

and

$$\theta_2 = 3000 \text{ h}$$

In this way the overall objective is achieved with the optimum design requirement being placed on each unit. The same philosophy should be applied to the allocation of repair times such that more attention is given to repair times in the high failure-rate areas.

System reliability and maintainability are not necessarily defined by a single MTBF and MTTR. It was emphasized in Chapter 2 that it is essential to treat each failure mode separately and, perhaps, to describe it by means of different parameters. For example, the requirement for an item of control equipment might be stated as follows:

- *Spurious failure* whereby a plant shutdown is carried out despite no valid shutdown condition:

    MTBF – 10 years

- *Failure to respond* whereby a valid shutdown condition does not lead to a plant shutdown (NB: a dormant failure):

    Probability of failure on demand which is, in fact, the unavailability = 0.0001

(NB: The unavailability is therefore 0.0001 and thus the availability is 0.9999. The MTBF is therefore determined by the down time since Unavailability is approximated from Failure Rate × Down Time.)

11.2 STRESS ANALYSIS

Component failure rates are very sensitive to the stresses applied. Stresses, which can be classified as environmental or self-generated, include:

Temperature
Shock
Vibration } Environmental
Humidity
Ingress of foreign bodies

Power dissipation
Applied voltage and current } Self-generated
Self-generated vibration
Wear

The sum of these stresses can be pictured as constantly varying, with peaks and troughs, and to be superimposed on a distribution of strength levels for a group of devices. A failure is seen as the result of stress exceeding strength. The average strength of the group of devices will increase during the early failures period owing to the elimination, from the population, of the weaker items.

Figure 11.2 Strength and stress

Random failures are assumed to result from the overlap of chance peaks in the stress distribution with the weaknesses in the population. It is for this reason that screening and burn-in are highly effective in decreasing component failure rates. During wearout, strength declines owing to physical and chemical processes. An overall change in the average stress will cause more of the peaks to exceed the strength values and more failures will result. Figure 11.2 illustrates this concept, showing a range of strength illustrated as a bold curve overlapping with a distribution of stress shown by the dotted curve. At the left-hand end of the diagram the strength is shown increasing as the burn-in failures are eliminated. Although not shown, wearout would be illustrated by the strength curves falling again at the right-hand end.

For specific stress parameters, calculations are carried out on the distributions of values. The devices in question can be tested to destruction in order to establish the range of strengths. The distribution of stresses is then obtained and the two compared. In Figure 11.2 the two curves are shown to overlap significantly in order to illustrate the concept, whereas in practice that overlap is likely to be at the extreme tails of two distributions. The data obtained may well describe the central shapes of each distribution but there is no guarantee that the tails will follow the model which has been assumed. The result would then be a wildly inaccurate estimate of the failure probability. The stress/strength concept is therefore a useful model to understand failure mechanisms, but only in particular cases can it be used to make quantitative predictions.

The principle of operating a component part below the rated stress level of a parameter in order to obtain a longer or more reliable life is well known. It is particularly effective in electronics where under-rating of voltage and temperature produces spectacular improvements in reliability. Stresses can be divided into two broad categories – environmental and operating.

Operating stresses are present when a device is active. Examples are voltage, current, self-generated temperature and self-induced vibration. These have a marked effect on the frequency of random failures as well as hastening wearout. Figure 11.3 shows the relationship of failure rate to the voltage and temperature stresses for a typical wet aluminium capacitor.

Figure 11.3

Note that a 5 to 1 improvement in failure rate is obtained by either a reduction in voltage stress from 0.9 to 0.3 or a 30 °C reduction in temperature. The relationship of failure rate to stress in electronic components is often described by a form of the Arrhenius equation which relates chemical reaction rate to temperature. Applied to random failure rate, the following two forms are often used:

$$\lambda_2 = \lambda_1 \exp K \left(\frac{1}{T_1} - \frac{1}{T_2} \right)$$

$$\lambda_2 = \lambda_1 \left(\frac{V_2}{V_1} \right)^n G^{(T_2 - T_1)}$$

V_2, V_1, T_2 and T_1 are voltage and temperature levels. λ_2 and λ_1 are failure rates at those levels. K, G and n are constants.

It is dangerous to use these types of empirical formulae outside the range over which they have been validated. Unpredicted physical or chemical effects may occur which render them inappropriate and the results, therefore, can be misleading. Mechanically, the principle of excess material is sometimes applied to increase the strength of an item. It must be remembered that this can sometimes have the reverse effect and the elimination of large sections in a structure can increase the strength and hence reliability.

A number of general derating rules have been developed for electronic items. They are summarized in the following table as percentages of the rated stress level of the component. In most cases two figures are quoted, these being the rating levels for High Reliability and Good Practice, respectively. The temperatures are for hermetic packages and 20°C should be deducted for plastic encapsulation.

	Maximum junction temp. (°C)	% of rated voltage	% of rated current	% of rated power	Fanout
Microelectronics					
– Linear	100/110	70/80	75/80		
– Hybrid	100				
– Digital TTL	120/130	75/85			75/80
– Digital MOS	100/105	75/85			75/80
Transistor					
– Si signal	110/115	60/80	75/85	50/75	
– Si power	125/130	60/80	60/80	30/50	
– FET junction	125	75/85		50/70	
– FET MOS	85/90	50/75		30/50	
Diode					
– Si signal	110/115	50/75	50/75	50/75	
– Si power/SCR	110/115	50/70	50/75	30/50	
– Zener	110/115		50/75	50/75	
Resistor					
– Comp. and Film				50/60	
– Wire wound				50/70	
Capacitor		40/50			
Switch and Relay contact					
– resistive/capacitive			70/75		
– inductive			30/40		
– rotating			10/20		

11.3 ENVIRONMENTAL STRESS PROTECTION

Environmental stress hastens the onset of wearout by contributing to physical deterioration. Included are:

Stress	Symptom	Action
High temperature	Insulation materials deteriorate. Chemical reactions accelerate	Dissipate heat. Minimize thermal contact. Use fins. Increase conductor sizes on PCBs. Provide conduction paths
Low temperature	Mechanical contraction damage. Insulation materials deteriorate	Apply heat and thermal insulation
Thermal shock	Mechanical damage within LSI components	Shielding
Mechanical shock	Component and connector damage	Mechanical design. Use of mountings
Vibration	Hastens wearout and causes connector failure	Mechanical design
Humidity	Coupled with temperature cycling causes 'pumping' – filling up with water	Sealing. Use of silica gel
Salt atmosphere	Corrosion and insulation degradation	Mechanical protection
Electromagnetic radiation	Interference to electrical signals	Shielding and part selection
Dust	Long-term degradation of insulation. Increased contact resistance	Sealing. Self-cleaning contacts
Biological effects	Decayed insulation material	Mechanical and chemical protection
Acoustic noise	Electrical interference due to microphonic effects	Mechanical buffers
Reactive gases	Corrosion of contacts	Physical seals

11.4 FAILURE MECHANISMS

11.4.1 Types of failure mechanism

The majority of failures are attributable to one of the following physical or chemical phenomena.

Alloy formation Formation of alloys between gold, aluminium and silicon causes what is known as 'purple plague' and 'black plague' in silicon devices.

Biological effects   Moulds and insects can cause failures. Tropical environments are particularly attractive for moulds and insects, and electronic devices and wiring can be affected.

Chemical and electrolytic changes   Electrolytic corrosion can occur wherever a potential difference together with an ionizable film are present. The electrolytic effect causes interaction between the salt ions and the metallic surfaces which act as electrodes. Salt-laden atmospheres cause corrosion of contacts and connectors. Chemical and physical changes to electrolytes and lubricants both lead to degradation failures.

Contamination   Dirt, particularly carbon or ferrous particles, causes electrical failure. The former deposited on insulation between conductors leads to breakdown and the latter to insulation breakdown and direct short circuits. Non-conducting material such as ash and fibrous waste can cause open-circuit failure in contacts.

Depolymerization   This is a degrading of insulation resistance caused by a type of liquefaction in synthetic materials.

Electrical contact failures   Failures of switch and relay contacts occur owing to weak springs, contact arcing, spark erosion and plating wear. In addition, failures due to contamination, as mentioned above, are possible. Printed-board connectors will fail owing to loss of contact pressure, mechanical wear from repeated insertions and contamination.

Evaporation   Filament devices age owing to evaporation of the filament molecules.

Fatigue   This is a physical/crystalline change in metals which leads to spring failure, fracture of structural members, etc.

Film deposition   All plugs, sockets, connectors and switches with non-precious metal surfaces are likely to form an oxide film which is a poor conductor. This film therefore leads to high-resistance failures unless a self-cleaning wiping action is used.

Friction   Friction is one of the most common causes of failure in motors, switches, gears, belts, styli, etc.

Ionization of gases   At normal atmospheric pressure a.c. voltages of approximately 300 V across gas bubbles in dielectrics give rise to ionization which causes both electrical noise and ultimate breakdown. This reduces to 200 V at low pressure.

Ion migration   If two silver surfaces are separated by a moisture-covered insulating material then, providing an ionizable salt is present as is usually the case, ion migration causes a silver 'tree' across the insulator.

Magnetic degradation   Modern magnetic materials are quite stable. However, degraded magnetic properties do occur as a result of mechanical vibration or strong a.c. electric fields.

Mechanical stresses   Bump and vibration stresses affect switches, insulators, fuse mountings, component lugs, printed-board tracks, etc.

Metallic effects   Metallic particles are a common cause of failure as mentioned above. Tin and cadmium can grow 'whiskers', leading to noise and low-resistance failures.

Moisture gain or loss   Moisture can enter equipment through pin holes by moisture vapour diffusion. This is accelerated by conditions of temperature cycling under high humidity. Loss of moisture by diffusion through seals in electrolytic capacitors causes reduced capacitance.

Molecular migration   Many liquids can diffuse through insulating plastics.

Stress relaxation   Cold flow ('creep') occurs in metallic parts and various dielectrics under mechanical stress. This leads to mechanical failure. This is not the same as fatigue which is caused by repeated movement (deformation) of a material.

Temperature cycling   This can be the cause of stress fluctuations, leading to fatigue or to moisture build-up.

11.4.2 Failures in semiconductor components

The majority of semiconductor device failures are attributable to the wafer-fabrication process. The tendency to create chips with ever-decreasing cross-sectional areas increases the probability that impurities, localized heating, flaws, etc., will lead to failure by deterioration, probably of the Arrhenius type (Section 11.2). Table 11.1 shows a typical proportion of failure modes.

Table 11.1

	Specific			In general (%)
	Linear (%)	TTL (%)	CMOS (%)	
Metallization	18	50	25	55
Diffusion	1	1	9	
Oxide	1	4	16	
Bond – die	10	10	–	25
Bond – wire	9	15	15	
Packaging/hermeticity	5	14	10	20
Surface contamination	55	5	25	
Cracked die	1	1	–	

11.4.3 Discrete components

The most likely causes of failure in resistors and capacitors are shown in Tables 11.2 and 11.3. Short-circuit failure is rare in resistors. For composition resistors, fixed and variable, the division tends to be 50% degradation failures and 50% open circuit. For film and wire-wound resistors the majority of failures are of the open-circuit type.

11.5 COMPLEXITY AND PARTS

11.5.1 Reduction of complexity

Higher scales of integration in electronic technology enable circuit functions previously requiring many hundreds (or thousands) of devices to be performed by a single component.

Table 11.2

Resistor type	Short	Open	Drift
Film	Insulation breakdown due to humidity. Protuberances of adjacent spirals	Mechanical breakdown of spiral due to r.f. Thin spiral	–
Wire wound	Over-voltage	Mechanical breakdown due to r.f. Failure of winding termination	
Composition			r.f. produces capacitance or dielectric loss
Variable (wire and composition)		Wiper arm wear. Excess current over a small segment owing to selecting low value	Noise

Mechanical movement |

Table 11.3

Capacitor type	Short	Open	Drift
Mica	Water absorption. Silver ion migration	Mechanical vibration	
Electrolytic solid tantalum	Solder balls caused by external heat from soldering	Internal connection. Failures due to shock or vibration	
Electrolytic non-solid tantalum	Electrolyte leakage due to temperature cycling	External welds	
Electrolytic aluminium oxide		Lead dissolved in electrolyte	Low capacitance due to aluminium oxide combining with electrolyte
Paper	Moisture. Rupture	Poor internal connections	
Plastic	Internal solder flow. Instantaneous breakdown in plastic causing s/c	Poor internal connections	
Ceramic	Silver ion migration	Mechanical stress. Heat rupture internal	
Air (variable)	Loose plates. Foreign bodies	Ruptured internal connections	

Hardware failure is restricted to either the device or its connections (sometimes 40 pins) to the remaining circuitry. A reduction in total device population and quantity leads, in general, to higher reliability.

Standard circuit configurations help to minimize component populations and allow the use of proven reliable circuits. Regular planned design reviews provide an opportunity to assess the economy of circuitry for the intended function. Digital circuits provide an opportunity for reduction in complexity by means of logical manipulation of the expressions involved. This enables fewer logic functions to be used in order to provide a given result.

11.5.2 Part selection

Since hardware reliability is largely determined by the component parts, their reliability and fitness for purpose cannot be over-emphasized. The choice often arises between standard parts with proven performance which just meet the requirement and special parts which are totally applicable but unproven. Consideration of design support services when selecting a component source may be of prime importance when the application is a new design. General considerations should be:

- Function needed and the environment in which it is to be used.
- Critical aspects of the part as, for example, limited life, procurement time, contribution to overall failure rate, cost, etc.
- Availability – number of different sources.
- Stress – given the application of the component the stresses applied to it and the expected failure rate. The effect of burn-in and screening on actual performance.

11.5.3 Redundancy

This involves the use of additional active units or of standby units. Reliability may be enhanced by this technique which can be applied in a variety of configurations:

- *Active Redundancy* – Full – With duplicated units, all operating, one surviving unit ensures non-failure.
  Partial – A specified number of the units may fail as, for example, two out of four engines on an aircraft. Majority voting systems often fall into this category.
  Conditional – A form of redundancy which occurs according to the failure mode.
- *Standby Redundancy* – Involves extra units which are not brought into use until the failure of the main unit is sensed.
- *Load Sharing* – Active redundancy where the failure of one unit places a greater stress on the remaining units.
- *Redundancy and Repair* – Where redundant units are subject to immediate or periodic repair, the system reliability is influenced both by the unit reliability and the repair times.

The decision to use redundancy must be based on an analysis of the trade-offs involved. It may prove to be the only available method when other techniques have been exhausted. Its application is not without penalties since it increases weight, space and cost and the increase in number of parts results in an increase in maintenance and spares holding costs. Remember, as we saw in Chapter 2, redundancy can increase the reliability for one failure mode but at the expense of another. In general, the reliability gain obtained from additional elements decreases beyond a few duplicated elements owing to either the common mode effects (Section 8.2) or to the reliability of devices needed to implement the particular configuration employed. Chapters 7 to 9 deal, in detail, with the quantitative effects of redundancy.

11.6 BURN-IN AND SCREENING

For an established design the early failures portion of the Bathtub Curve represents the populations of items having inherent weaknesses due to minute variations and defects in the manufacturing process. Furthermore, it is increasingly held that electronic failures – even in the constant failure rate part of the curve – are due to microscopic defects in the physical build of the item. The effects of physical and chemical processes with time cause failures to occur both in the early failures and constant failure rate portions of the Bathtub. Burn-in and Screening are thus effective means of enhancing component reliability:

- *Burn-in* is the process of operating items at elevated stress levels (particularly temperature, humidity and voltage) in order to accelerate the processes leading to failure. The populations of defective items are thus reduced.
- *Screening* is an enhancement to Quality Control whereby additional detailed visual and electrical/mechanical tests seek to reveal defective features which would otherwise increase the population of 'weak' items.

The relationship between various defined levels of Burn-in and Screening and the eventual failure rate levels is recognized and has, in the case of electronic components, become formalized. For microelectronic devices US MIL STD 883 provides a uniform set of Test, Screening and Burn-in procedures. These include tests for moisture resistance, high temperature, shock, dimensions, electrical load and so on. The effect is to eliminate the defective items mentioned above. The tests are graded into three classes in order to take account of the need for different reliability requirements at appropriate cost levels. These levels are:

Class C – the least stringent which requires 100% internal visual inspection. There are electrical tests at 25°C but no Burn-in.

Class B – in addition to the requirements of Class C there is 160 hours of Burn-in at 125°C and electrical tests at temperature extremes (high and low).

Class S – in addition to the tests in Class B there is longer Burn-in (240 hours) and more severe tests including 72 hours reverse bias at 150°C.

The overall standardization and QA programmes described in US-MIL-M-38510 call for the MIL 883 tests procedures. The UK counterpart to the system of controls is BS 9000, which functions as a four-tier hierarchy of specifications from the general requirements at the top, through generic requirements, to detail component manufacture and test details at the bottom. Approximate equivalents for the screening levels are:

MIL 883	BS 9400	Relative cost (approx.)
S	A	10
B	B	5
C	C	3
–	D	1
		0.5 (plastic)

11.7 MAINTENANCE STRATEGIES

This is dealt with, under reliability centred maintenance, in Chapter 16. It involves:

- Routine maintenance (adjustment, overhaul)
- Preventive discard (replacement)
- Condition monitoring (identifying degradation)
- Proof testing for dormant redundant failures

12 Design review and test

12.1 REVIEW TECHNIQUES

Design review is the process of comparing the design, at points during design and development, with the requirements of earlier stages. Examples are a review of:

- The functional specification against the requirements specification;
- Circuit or mechanical assembly performance against the functional specification;
- Predicted reliability/availability against targets in the requirements specification;
- Some software source code against the software specification.

Two common misconceptions about design review are:

- That they are schedule progress meetings;
- That they are to appraise the designer.

They are, in fact, to verify the design, as it exists at a particular time against the requirements. It is a measure, as is test, but carried out by document review and predictive calculations. The results of tests may well be an input to the review but the review itself is an intellectual process rather than a test.

It is a feedback loop which verifies each stage of design and provides confidence to proceed to the next. Review is a formal activity and should not be carried out casually. The following points are therefore important when conducting reviews:

- They must be carried out against a defined baseline of documents. In other words, the design must be frozen at specific points, known as baselines, which are defined by a list of documents and drawings each at a specific issue status.
- Results must be recorded and remedial actions formally followed up.
- All documents must be available in advance and checklists prepared of the points to be reviewed.
- Functions and responsibilities of the participants must be defined.
- The review must be chaired by a person independent of the design.
- The purpose must be specific and stated in advance in terms of what is to be measured. Consequently, the expected results should be laid down.

The natural points in the design cycle which lend themselves to review are:

1. *Requirements specification*: This is the one point in the design cycle above which there is no higher specification against which to compare. It is thus the hardest review in terms of deciding if the outcome is satisfactory. Nevertheless, features such as completeness,

unambiguity and consistency can be considered. A requirement specification should not prejudge the design and therefore it can be checked that it states *what* is required rather than *how* it is to be achieved.

2. *Functional specification*: This can be reviewed against the requirements specification and each function checked off for accuracy or omission.
3. *Detailed design*: This may involve a number of reviews depending on how many detailed design documents/modules are created. At this level, horizontal, as well as vertical, considerations arise. In addition to measuring the documents' compliance with the preceding stages it is necessary to examine its links to other specifications/modules/drawings/diagrams, etc. Reliability predictions and risk assessments, as well as early test results, are used as inputs to measure the assessed conformance to higher requirements.
4. *Software*: Code reviews are a particular type of review and are covered in Section 17.4.5.
5. *Test results*: Although test follows later in the design cycle, it too can be the subject of review. It is necessary to review the test specifications against the design documents (e.g. functional specification). Test results can also be reviewed against the test specification.

A feature of review is the checklist. This provides some structure for the review and can be used for recording the results. Also, checklists are a means of adding questions based on experience and can be evolved as lessons are learned from reviews. Section 17.6 provides specific checklists for software reviews. It is important, however, not to allow checklists to constrain the review process since they are only an *aide-mémoire*.

12.2 CATEGORIES OF TESTING

There are four categories of testing:

1. *Design Testing* – Laboratory and prototype tests aimed at proving that a design will meet the specification. Initially prototype functional tests aim at proving the design. This will extend to pre-production models which undergo environmental and reliability tests and may overlap with:
2. *Qualification Testing* – Total proving cycle using production models over the full range of the environmental and functional specification. This involves extensive marginal tests, climatic and shock tests, reliability and maintainability tests and the accumulation of some field data. It must not be confused with development or production testing. The purpose of Qualification Testing is to ensure that a product meets all the requirements laid down in the Engineering Specification. This should not be confused with product testing which takes place after manufacture. Items to be verified are:

Function – Specified performance at defined limits and margins.
Environment – Ambient Temperature and Humidity for use, storage, etc. Performance at the extremes of the specified environment should be included.
Life – At specified performance levels and under storage conditions.
Reliability – Observed MTBF under all conditions.
Maintainability – MTTR/MDT for defined test equipment, spares, manual and staff.
Maintenance – Is the routine and corrective maintenance requirement compatible with use?
Packaging and Transport – Test under real conditions including shock tests.
Physical characteristics – Size, weight, power consumption, etc.

Ergonomics – Consider interface with operators and maintenance personnel.

Testability – Consider test equipment and time required for production models.

Safety – Use an approved test house such as BSI or the British Electrotechnical Approvals Board.

3. *Production Testing and Commissioning* – Verification of conformance by testing modules and complete equipment. Some reliability proving and burn-in may be involved. Generally, failures will be attributable to component procurement, production methods, etc. Design-related queries will arise but should diminish in quantity as production continues.

4. *Demonstration Testing* – An acceptance test whereby equipment is tested to agreed criteria and passes or fails according to the number of failures.

These involve the following types of test.

12.2.1 Environmental testing

This proves that equipment functions to specification (for a sustained period) and is not degraded or damaged by defined extremes of its environment. The test can cover a wide range of parameters and it is important to agree a specification which is realistic. It is tempting, when in doubt, to widen the limits of temperature, humidity and shock in order to be extra sure of covering the likely range which the equipment will experience. The resulting cost of over-design, even for a few degrees of temperature, may be totally unjustified.

The possibilities are numerous and include:

Electrical

Electric fields.

Magnetic fields.

Radiation.

Climatic

Temperature extremes ⎫
Temperature cycling ⎬ internal and external may be specified.
                    ⎭

Humidity extremes.

Temperature cycling at high humidity.

Thermal shock – rapid change of temperature.

Wind – both physical force and cooling effect.

Wind and precipitation.

Direct sunlight.

Atmospheric pressure extremes.

Mechanical

Vibration at given frequency – a resonant search is often carried out.

Vibration at simultaneous random frequencies – used because resonances at different frequencies can occur simultaneously.

Mechanical shock – bump.

Acceleration.

Chemical and hazardous atmospheres

Corrosive atmosphere – covers acids, alkalis, salt, greases, etc.

Foreign bodies – ferrous, carbon, silicate, general dust, etc.

Biological – defined growth or insect infestation.

Reactive gases.

Flammable atmospheres.

12.2.2 Marginal testing

This involves proving the various system functions at the extreme limits of the electrical and mechanical parameters and includes:

Electrical
Mains supply voltage.
Mains supply frequency.
Insulation limits.
Earth testing.
High voltage interference – radiated. Typical test apparatus consists of a spark plug, induction coil and break contact.
Mains-borne interference.
Line error rate – refers to the incidence of binary bits being incorrectly transmitted in a digital system. Usually expressed as in 1 in 10^{-n} bits.
Line noise tests – analogue circuits.
Electrostatic discharge – e.g. 10 kV from 150 pF through 150 Ω to conductive surfaces.
Functional load tests – loading a system with artificial traffic to simulate full utilization (e.g. call traffic simulation in a telephone exchange).
Input/output signal limits – limits of frequency and power.
Output load limits – sustained voltage at maximum load current and testing that current does not increase even if load is increased as far as a short circuit.
Mechanical
Dimensional limits – maximum and minimum limits as per drawing.
Pressure limits – covers hydraulic and pneumatic systems.
Load – compressive and tensile forces and torque.

12.2.3 High-reliability testing

The major problem in verifying high reliability, emphasized in Chapter 5, is the difficulty of accumulating sufficient data, even with no failures, to demonstrate statistically the value required. If an MTBF of, say, 10^6 h is to be verified, and 500 items are available for test, then 2000 elapsed hours of testing (3 months of continuous test) are required to accumulate sufficient time for even the minimum test which involves no failures. In this way, the MTBF is demonstrated with 63% confidence. Nearly two and a half times the amount of testing is required to raise this to 90%.

The usual response to this problem is to acclerate the failure mechanisms by increasing the stress levels. This involves the assumption that relationships between failure rate and stress levels hold good over the range in question. Interpolation between points in a known range presents little problem whereas extrapolation beyond a known relationship is of dubious value. Experimental data can be used to derive the constants found in the equations shown in Section 11.2. In order to establish if the Arrhenius relationship applies, a plot of $\log_e$ failure rate against the reciprocal of temperature is made. A straight line indicates that it holds for the temperature range in question. In some cases parameters such as ambient temperature and power are not independent, as in transistors where the junction temperature is a function of both. Accelerated testing gives a high confidence that the failure rate at normal stress levels is, at least, less than that observed at the elevated stresses.

Where MTBF is expressed in cycles or operations, as with relays, pistons, rods and cams, the test may be accelerated without a change in the physics of the failure mechanism. For example, 100 contactors can be operated to accumulate 3×10^8 operations in one month although, in normal use, it might well take several years to accumulate the same amount of data.

12.2.4 Testing for packaging and transport

There is little virtue in investing large sums in design and manufacture if inherently reliable products are to be damaged by inadequate packaging and handling. The packaging needs to match the characteristics and weaknesses of the contents with the hazards it is likely to meet. The major causes of defects during packaging, storage and transport are:

1. Inadequate or unsuitable packaging materials for the transport involved.
   Transport, climatic and vibration conditions not foreseen.
   Storage conditions and handling not foreseen.
   – requires consideration of waterproofing, hoops, bands, lagging, hermetic seals, desiccant, ventilation holes, etc.
2. Inadequate marking – see BS 2770 Pictorial handling instructions.
3. Failure to treat for prevention of corrosion.
   – various cleaning methods for the removal of oil, rust and miscellaneous contamination followed by preventive treatments and coatings.
4. Degradation of packaging materials owing to method of storage prior to use.
5. Inadequate adjustments or padding prior to packaging.
   Lack of handling care during transport.
   – requires adequate work instructions, packing lists, training, etc.

Choosing the most appropriate packaging involves considerations of cost, availability and size, for which reason a compromise is usually sought. Crates, rigid and collapsible boxes, cartons, wallets, tri-wall wrapping, chipboard cases, sealed wrapping, fabricated and moulded spacers, corner blocks and cushions, bubble wrapping, etc. are a few of the many alternatives available to meet any particular packaging specification.

Environmental testing involving vibration and shock tests together with climatic tests is necessary to qualify a packaging arrangement. This work is undertaken by a number of test houses and may save large sums if it ultimately prevents damaged goods being received since the cost of defects rises tenfold and more, once equipment has left the factory. As well as specified environmental tests, the product should be transported over a range of typical journeys and then retested to assess the effectiveness of the proposed pack.

12.2.5 Multiparameter testing

More often than not, the number of separate (but not independent) variables involved in a test makes it impossible for the effect of each to be individually assessed. To hold, in turn, all but one parameter constant and record its effect and then to analyse and relate all the parametric results would be very expensive in terms of test and analysis time. In any case, this has the drawback of restricting the field of data. Imagine that, in a three-variable situation, the limits are represented by the corners of a cube as in Figure 12.1, then each test would be confined to a straight line through the cube.

Figure 12.1

One effective approach involves making measurements of the system performance at various points, including the limits, of the cube. For example, in a facsimile transmission system the three variables might be the line error rate, line bandwidth and degree of data compression. For each combination the system parameters would be character error rate on received copy and transmission time. Analysis of the cube would reveal the best combination of results and system parameters for a cost-effective solution.

12.2.6 Step-stress testing

This can involve increasing one or more parameters. The stress parameters chosen (e.g. temperature, mechanical load) are increased by increments at defined time intervals. Thus, for example, a mechanical component could be tested at its nominal temperature and loading for a period of time. Both temperature and load would then be increased by a defined amount for a further equal period. Successive increments of stress would then be applied after each period.

The median rank cumulative failure percentages would then be plotted against the failure times (loglog against log) and a line obtained which (assuming the majority of failures occurred at the higher stresses) can be extrapolated back to the normal stress condition. The target probability of failure for some defined time period, at normal stress, will be a single point on the graph paper.

If the target point falls well to the left of the line then there is SOME evidence that the design is adequate. Advantages and disadvantages of such a judgement are:

ADVANTAGES:
   Gives some indication of failure free life
   Gives some confidence in the design

DISADVANTAGES
   The assumption of linearity of the plot may not be valid
   Does not address all combinations of stresses
   Inaccuracies in the plot

12.3 RELIABILITY GROWTH MODELLING

This concerns the improvement in reliability, during use, which comes from field data feedback resulting in modifications. Improvements depend on ensuring that field data actually lead to design modifications. Reliability growth, then, is the process of eliminating design-related failures. It must not be confused with the decreasing failure rate described by the Bathtub Curve.

Figure 12.2

Figure 12.2 illustrates this point by showing two Bathtub Curves for the same item of equipment. Both show an early decreasing failure rate whereas the later model, owing to reliability growth, shows higher Reliability in the random failures part of the curve.

A simple but powerful method of plotting growth is the use of CUSUM (Cumulative Sum Chart) plots. In this technique an anticipated target MTBF is chosen and the deviations are plotted against time. The effect is to show the MTBF by the slope of the plot, which is more sensitive to changes in Reliability.

The following example shows the number of failures after each 100 h of running of a generator. The CUSUM is plotted in Figure 12.3.

Cumulative hours	Failures	Anticipated failures if MTBF were 200 h	Deviation	CUSUM
100	1	0.5	+0.5	+0.5
200	1	0.5	+0.5	+1
300	2	0.5	+1.5	+2.5
400	1	0.5	+0.5	+3
500	0	0.5	−0.5	+2.5
600	1	0.5	+0.5	+3
700	0	0.5	−0.5	+2.5
800	0	0.5	−0.5	+2
900	0	0.5	−0.5	+1.5
1000	0	0.5	−0.5	+1

Figure 12.3 CUSUM plot

The CUSUM is plotted for an objective MTBF of 200 h. It shows that for the first 400 h the MTBF was in the order of half the requirement. From 400 to 600 h there was an improvement to about 200 h MTBF and thereafter there is evidence of reliability growth. The plot is sensitive to the changes in trend, as can be seen from the above.

The reader will note that the axis of the deviation has been inverted so that negative variations produce an upward trend. This is often done in reliability CUSUM work in order to reflect improving MTBFs by an upward curve, and vice versa.

Whereas CUSUM provides a clear picture of past events, it is sometimes required to establish a relationship between MTBF and time for the purposes of predicting Reliability Growth. The best-known model is that described by J. T. DUANE, in 1962. It assumes an empirical relationship whereby the improvement in MTBF is proportional to T^α where T is the total equipment time and α is a growth factor.

This can be expressed in the form:

$$\theta = k\, T^\alpha$$

Which means that:

$$\theta_2/\theta_{21} = (T_2/T_1)^\alpha$$

Hence, if any two values of T and MTBF are known the equations can be solved to obtain k and α. The amount of T required to reach a given desired MTBF can then be predicted, with the assumption that the growth rate does not change.

Typically α is between 0.1 and 0.65. Figure 12.4 shows a Duane plot of cumulative MTBF against cumulative time on log axes. The range, r, shows the cumulative times required to achieve a specific MTBF for factors between 0.2 and 0.5.

A drawback to the Duane plot is that it does not readily show changes in the growth rate since the data are effectively smoothed. This effect becomes more pronounced as the plot progresses since, by using cumulative time, any sudden deviations are damped.

It is a useful technique during a field trial for predicting, at the current growth rate, how many field hours need to be accumulated in order to reach some target MTBF. In Figure 12.4, if the $\alpha = 0.2$ line was obtained from field data after, say, 800 cumulative field years then, if the objective MTBF were 500 years, the indication is that 10 000 cumulative years are needed at that growth rate. The alternative would be to accelerate the reliability growth by more active follow-up of the failure analysis.

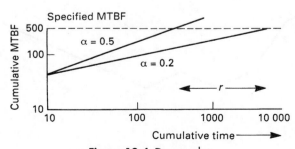

Figure 12.4 Duane plot

EXERCISES

1. 100 items are placed on simulated life test. Failures occur at:

   17, 37, 45, 81, 88, 110, 122, 147, 208, 232, 235, 263, 272, 317, 325, 354, 355, 403 hours.
   A 3000 hour MTBF is hoped for. Construct a CUSUM, in 3000 cumulative hour increments, to display these results.

2. 50 items are put on field trial for 3 months and have generated 20 failures.
   A further 50 are added to the trial and, after a further 3 months, the total number of failures has risen to 35.
   Construct a Duane plot, on log/log paper, and determine when the MTBF will reach 12 000 hours?
   Calculate the growth factor.
   If the growth factor is increased to 0.6 when will an MTBF of 12 000 hours be reached?

13 Field data collection and feedback

13.1 REASONS FOR DATA COLLECTION

Failure data can be collected from prototype and production models or from the field. In either case a formal failure-reporting document is necessary in order to ensure that the feedback is both consistent and adequate. Field information is far more valuable since it concerns failures and repair actions which have taken place under real conditions. Since recording field incidents relies on people, it is subject to errors, omissions and misinterpretation. It is therefore important to collect all field data using a formal document. Information of this type has a number of uses, the main two being feedback, resulting in modifications to prevent further defects, and the acquisition of statistical reliability and repair data. In detail, then, they:

- Indicate design and manufacture deficiencies and can be used to support reliability growth programmes (Section 12.3);
- Provide quality and reliability trends;
- Identify wearout and decreasing failure rates;
- Provide subcontractor ratings;
- Contribute statistical data for future reliability and repair time predictions;
- Assist second-line maintenance (workshop);
- Enable spares provisioning to be refined;
- Allow routine maintenance intervals to be revised;
- Enable the field element of quality costs to be identified.

A failure-reporting system should be established for every project and product. Customer cooperation with a reporting system is essential if feedback from the field is required and this could well be sought, at the contract stage, in return for some other concession.

13.2 INFORMATION AND DIFFICULTIES

A failure report form must collect information covering the following:

- Repair time – active and passive.
- Type of fault – primary or secondary, random or induced, etc.
- Nature of fault – open or short circuit, drift condition, wearout, design deficiency.
- Fault location – exact position and detail of LRA or component.
- Environmental conditions – where these are variable, record conditions at time of fault if possible.

- Action taken – exact nature of replacement or repair.
- Personnel involved.
- Equipment used.
- Spares used.
- Unit running time.

The main problems associated with failure recording are:

1. *Inventories*: Whilst failure reports identify the numbers and types of failure they rarely provide a source of information as to the total numbers of the item in question and their installation dates and running times.
2. *Motivation*: If the field service engineer can see no purpose in recording information it is likely that items will be either omitted or incorrectly recorded. The purpose of fault reporting and the ways in which it can be used to simplify the task need to be explained. If the engineer is frustrated by unrealistic time standards, poor working conditions and inadequate instructions, then the failure report is the first task which will be skimped or omitted. A regular circulation of field data summaries to the field engineer is the best (possibly the only) way of encouraging feedback. It will help him to see the overall field picture and advice on diagnosing the more awkward faults will be appreciated.
3. *Verification*: Once the failure report has left the person who completes it the possibility of subsequent checking is remote. If repair times or diagnoses are suspect then it is likely that they will go undetected or be unverified. Where failure data are obtained from customer's staff, the possibility of challenging information becomes even more remote.
4. *Cost*: Failure reporting is costly in terms of both the time to complete failure-report forms and the hours of interpretation of the information. For this reason, both supplier and customer are often reluctant to agree to a comprehensive reporting system. If the information is correctly interpreted and design or manufacturing action taken to remove failure sources, then the cost of the activity is likely to be offset by the savings and the idea must be 'sold' on this basis.
5. *Recording non-failures*: The situation arises where a failure is recorded although none exists. This can occur in two ways. First, there is the habit of locating faults by replacing suspect but not necessarily failed components. When the fault disappears the first (wrongly removed) component is not replaced and is hence recorded as a failure. Failure rate data are therefore artificially inflated and spares depleted. Second, there is the interpretation of secondary failures as primary failures. A failed component may cause stress conditions upon another which may, as a result, fail. Diagnosis may reveal both failures but not always which one occurred first. Again, failure rates become wrongly inflated. More complex maintenance instructions and the use of higher-grade personnel will help reduce these problems at a cost.
6. *Times to failure*: These are necessary in order to establish wearout. See next section.

13.3 TIMES TO FAILURE

In most cases fault data schemes yield the numbers of failures/defects of equipment. Establishing the inventories, and the installation dates of items, is also necessary if the cumulative times are also to be determined. This is not always easy as plant records are often incomplete (or out of date) and the exact installation dates of items has sometimes to be guessed.

Nevertheless, establishing the number of failures and the cumulative time enables failure rates to be inferred as was described in Chapter 5.

Although this failure rate information provides a valuable input to reliability prediction and to optimum spares provisioning (Chapter 16), it does not enable the wearout and burn-in characteristics of an item to be described. In Chapter 6 the Weibull methodology for describing variable failure rates was described and, in Chapter 16 it is shown how to use this information to optimize replacement intervals.

For this to happen it is essential that each item is separately identified (usually by a tag number) and that each failure is attributed to a specific item. Weibull models are usually, although not always, applicable at the level of a specific failure mode rather than to the failures as a whole. A description of failure mode is therefore important and the physical mechanism, rather than the outcome, should be described. For example the phrase 'out of adjustment' really describes the effect of a failure whereas 'replaced leaking diaphragm' more specifically describes the mode.

Furthermore, if an item is removed, replaced or refurbished as new then this needs to be identified (by tag number) in order for the correct start times to be identified for each subsequent failure time. In other words if an item which has been in situ for 5 years had a new diaphragm fitted 1 year ago then, for diaphragm failures, the time to failure dates from the latter. On the other hand failures of another mode might well be treated as times dating from the former.

Another complication is in the use of operating time rather than calendar time. In some ways the latter is more convenient if the data is to be used for generic use. In some cases however, especially where the mode is related to wear and the operating time is short compared with calendar time, then operating hours will be more meaningful. In any case consistency is the rule.

If this information is available then it will be possible to list:

- individual times to failure (calendar or operating)
- times for items which did not fail
- times for items which were removed without failing

In summary the following are needed:

- Installed (or replaced/refurbished) dates and tag numbers
- Failure dates and tag numbers
- Failure modes (by physical failure mechanism)
- Running times/profiles unless calendar time is be used

13.4 SPREADSHEETS AND DATABASES

Many data-collection schemes arrange for the data to be manually transferred, from the written form, into a computer. In order to facilitate data sorting and analysis it is very useful if the information can be in a coded form. This requires some form of codes database for the field maintenance personnel in order that the various entries can be made by means of simple alphanumerics. This has the advantage that field reports are more likely to be complete since there is a code available for each box on the form. Furthermore, the codes then provide definitive classifications for subsequent sorting. Headings include:

Equipment code

Preferably a hierarchical coding scheme which defines the plant, subsystem and item as, for example, RC1-66-03-5555, where:

Code	Meaning
R	Southampton Plant
C1	Compression system
66	Power generation
03	Switchgear
5555	Actual item

How found

The reason for the defect being discovered as, say, a two-digit code:

Code	Meaning
01	Plant shutdown
02	Preventive maintenance
03	Operating problem
etc.	

Type of fault

The failure mode, for example:

Code	Meaning
01	Short circuit
02	Open circuit
03	Leak
04	Drift
05	No fault found
etc.	

Action taken

Examples are:

Code	Meaning
01	Item replaced
02	Adjusted
03	Item repaired
etc.	

Discipline

Where more than one type of maintenance skill is used, as is often the case on big sites, it is desirable to record the maintenance discipline involved. These are useful data for future maintenance planning and costing. Thus.

Code	Meaning
01	Electrical
02	Instrument
03	Mechanical
etc.	

Free text
In addition to the coded report there needs to be some provision for free text in order to amplify the data.

Each of the above fields may run to several dozen codes which would be issued to the field maintenance personnel as a handbook. Two suitable types of package for analysis of the data are *spreadsheets and databases*. If the data can be inputted directly into one of these packages, so much the better. In some cases the data are resident in a more wide-ranging, field-specific, computerized maintenance system. In those cases it will be worth writing a download program to copy the defect data into one of the above types of package.

Spreadsheets such as Lotus 1-2-3 and XL allow the data, including text, to be placed in cells arranged in rows and columns. Sorting is available as well as mathematical manipulation of the data.

In some cases the quantity of data may be such that spreadsheet manipulation becomes slow and cumbersome, or is limited by the extent of the PC memory. The use of *database* packages permits more data to be handled and more flexible and fast sorting. Sorting is far more flexible than with spreadsheets since words within text, within headings or even 'sound-alike' words can be sorted.

13.5 BEST PRACTICE AND RECOMMENDATIONS

The following list summarizes the best practice together with recommended enhancements for both manual and computer based field failure recording.

Recorded field information is frequently inadequate and it is necessary to emphasize that failure data must contain sufficient information to enable precise failures to be identified and failure distributions to be identified. They must, therefore, include:

(a) Adequate information about the symptoms and causes of failure. This is important because predictions are only meaningful when a system level failure is precisely defined. Thus component failures which contribute to a defined system failure can only be identified if the failure modes are accurately recorded. There needs to be a distinction between failures (which cause loss of system function) and defects (which may only cause degradation of function).

(b) Detailed and accurate equipment inventories enabling each component item to be separately identified. This is essential in providing cumulative operating times for the calculation of assumed constant failure rates and also for obtaining individual calendar times (or operating times or cycles) to each mode of failure and for each component item. These individual times to failure are necessary if failure distributions are to be analysed by the Weibull method dealt with in Chapter 6.

(c) Identification of common cause failures by requiring the inspection of redundant units to ascertain if failures have occurred in both (or all) units. This will provide data to enhance models such as the one developed in Chapter 8.2. In order to achieve this it is necessary to be able to identify that two or more failures are related to specific field items in a redundant configuration. It is therefore important that each recorded failure also identifies which specific item (i.e. tag number) it refers to.

(d) Intervals between common cause failures. Because common cause failures do not necessarily occur at precisely the same instant it is desirable to be able to identify the time elapsed between them.

(e) The effect that a 'component part' level failure has on failure at the system level. This will vary according to the type of system, the level of redundancy (which may postpone system level failure) etc.

(f) Costs of failure such as the penalty cost of system outage (e.g. loss of production) and the cost of corrective repair effort and associated spares and other maintenance costs.

(g) The consequences in the case of safety-related failures (e.g. death, injury, environmental damage) not so easily quantified.

(h) Consideration of whether a failure is intrinsic to the item in question or was caused by an external factor. External factors might include:

> process operator error induced failure
> maintenance error induced failure
> failure caused by a diagnostic replacement attempt
> modification induced failure

(i) Effective data screening to identify and correct errors and to ensure consistency. There is a cost issue here in that effective data screening requires significant man-hours to study the field failure returns. In the author's experience an average of as much as one hour per field return can be needed to enquire into the nature of a given failure and to discuss and establish the underlying cause. Both codification and narrative are helpful to the analyst and, whilst each has its own merits, a combination is required in practice. Modern computerized maintenance management systems offer possibilities for classification and codification of failure modes and causes. However, this relies on motivated and trained field technicians to input accurate and complete data. The option to add narrative should always be available.

(j) Adequate information about the environment (e.g. weather in the case of unprotected equipment) and operating conditions (e.g. unusual production throughput loadings).

13.6 ANALYSIS AND PRESENTATION OF RESULTS

Once collected, data must be analysed and put to use or the system of collection will lose credibility and, in any case, the cost will have been wasted. A Pareto analysis of defects is a powerful method of focusing attention on the major problems. If the frequency of each defect type is totalled and the types then ranked in descending order of frequency it will usually be seen that a high percentage of the defects are spread across only a few types. A still more useful approach, if cost information is available, is to multiply each defect type frequency by its cost and then to rerank the categories in descending order of cost. Thus the most expensive group of defects, rather than the most frequent, heads the list, as can be seen in Figure 13.1.

Note the emphasis on cost and that the total has been shown as a percentage of sales. It is clear that engineering effort could profitably be directed at the first two items which together account for 38% of the failure cost. The first item is a mechanical design problem and the second a question of circuit tolerancing.

It is also useful to know whether the failure rate of a particular failure type is increasing, decreasing or constant. This will influence the engineering response. A decreasing failure rate indicates the need for further action in tests to eliminate the early failures. Increasing failure rate shows wearout, requiring either a design solution or preventive replacement. Constant failure rate suggests a reliability level which is inherent to that design configuration. Chapter

1. *Summary of Data*

Number of machines in field	50
Operating hours (this period)	5320
Number of corrective calls	39
Total cost of calls	£4250 labour, travel and spares
Total cost as % of sales	4%

2. *Incident Analysis*

Repetitive Failures	Frequency	Cost £	% of total
a) Mechanical transporter assembly — belt adjustment	4	935	22
b) Receiver carrier detector drift	9	680	16
c) Electromechanical relays	4	340	8
d) Gear Meshing	3	340	8
e) Printed Board 182c output VT2	2	300	7
f) Lamps	2	170	4
Non-repetitive Faults			
g) Printed Board 424a IC5 } h) Printed Board 111e R2 }	15	1485	35
etc.	39	4250	100

Figure 13.1 Quarterly incident report summary - product Y

6 explains how failure data can be analysed to quantify these trends. The report in Figure 13.1 might well contain other sections showing reliability growth, analysis of wearout, progress on engineering actions since the previous report, etc.

13.7 EXAMPLES OF FAILURE REPORT FORMS

Although very old, Figure 13.2 shows an example of a well-designed and thorough failure recording form as once used by the European companies of the International Telephone and Telegraph Corporation. This single form strikes a balance between the need for detailed failure information and the requirement for a simple reporting format. A feature of the ITT form is the use of four identical print-through forms. The information is therefore accurately recorded four times with minimum effort.

   Figure 13.3 shows the author's recommended format taking into account the list of items in Section 13.5.

	System	Sub-system	Module/sub-assembly	
ITTE Failure Report and Action Form				

Type | | | | Report number:
Serial number | | | | Report date:
Location/identification | | | | Report completed by:
On-time (cumulative) | | | | Company:
Down time (this failure) | | | |
Active repair time | | | |

System status: — Details of symptoms, diagnosis and failure:
    Field service
    Field trial
    Production prototype
    Model

Effect of failure on system:
    Complete system failure
    Major degradation
    Minor degradation
    None

On-site diagnosis:
    No defect found
    Part failure
    Installation defect
    Manufacturing defect
    Design defect
    Program defect
    On-site human error
    Other

Action taken: — Details of action taken:
    Replace module
    Repair
    Modification
    Program reload
    Other

(Left margin: To be completed on site)

Project engineering action:	Name	Company/Dept	Signature	Date completed
– Consolidate with filed data				
– For immediate analysis/action by				
Engineering				
Manufacturing				
Quality assurance				
Purchasing				
Other				

Analysis and action taken:
    Engineering change no.:
    Dated:
    Follow-up report. Ref no.:
    Dated:
    Name:
    Signature:
    Date:

For information to:

(Left margin: To be completed at Designated Centre)

Figure 13.2 ITT Europe failure report and action form

SERIAL NUMBER ....

DATE (and time) OF INCIDENT/EVENT/FAILURE

DATE ITEM INSTALLED (or replaced or refurbished)

MAINTENANCE TECHNICIAN (Provides traceability)

DISCIPLINE (e.g, Electrical, Mechanical, Instrumentation)

FAILED COMPONENT ITEM DESCRIPTION (e.g, Motor)

SUBSYSTEM (e.g, Support system)

DESCRIPTION OF FAULT/CAUSE (Failure mode, e.g, Windings open circuit)

'TAG', 'SERIAL NUMBER' (HENCE DATE OF INSTALLATION AND REFURB)
e.g, System xyz, Unit abc, Motor type zzz, serial no. def,

DOWN TIME [if known]/ REPAIR TIME
e.g, 4 hrs repair, 24 hrs outage

TIME TO FAILURE (COMPUTED FROM DATE AND TAG NUMBER)
e.g, This date minus date of installation
e.g, This date minus date of last refurbishment

PARTS USED (in the repair)
e.g, New motor type zzz, serial number efg

ACTION TAKEN (e.g, Replace motor)

HOW CAUSED
Intrinsic (e.g, RANDOM HARDWARE FAILURE) versus extrinsic (GIVE CAUSE IF EVIDENT)

HOW FOUND/DIAGNOSED
e.g, Customer report, technician discovered open circuit windings

RESULT OF FAILURE ON SYSTEM
e.g, Support system un-usable, process trip, no effect

COMMON CAUSE FAILURE e.g, redundancy defeated
time between CCFs ....
attributable to SEPARATION/DIVERSITY/COMPLEXITY/HUMAN
FACTOR/ENVIRONMENT

ENVIRONMENT/OPERATING CONDITION
e.g, temp, humidity, 50% throughput, equipment unattended

NARRATIVE ....

Figure 13.3 Recommended failure data recording form

14 Factors influencing down time

The two main factors governing down time are equipment design and maintenance philosophy. In general, it is the active repair elements that are determined by the design and the passive elements which are governed by the maintenance philosophy. Designers must be aware of the maintenance strategy and of the possible equipment failure modes. They must understand that production difficulties can often become field problems since, if assembly is difficult, maintenance will be well-nigh impossible. Achieving acceptable repair times involves simplifying diagnosis and repair.

14.1 KEY DESIGN AREAS

14.1.1 Access

Low-reliability parts should be the most accessible and must be easily removable with the minimum of disturbance. There must be enough room to withdraw such devices without touching or damaging other parts. On the other hand, the technician must be discouraged from removing and checking easily exchanged items as a substitute for the correct diagnostic procedure. The use of captive screws and fasteners is highly desirable as they are faster to use and eliminate the risk of losing screws in the equipment. Standard fasteners and covers become familiar and hence easier to use. The use of outriggers, which enables printed boards to be tested while still electrically connected to the system, can help to reduce diagnosis time. On the other hand, this type of on-line diagnosis can induce faults and is sometimes discouraged. In general, it is a good thing to minimize on-line testing by employing easily interchanged units together with alarms and displays providing diagnostic information and easy identification of the faulty unit.

Every LRA (Least Replaceable Assembly) should be capable of removal without removing any other LRA or part. The size of the LRA affects the speed of access. The overall aim is for speedy access consistent with minimum risk of accidental damage.

14.1.2 Adjustment

The amount of adjustment required during normal system operation, and after LRA replacement, can be minimized (or eliminated) by generous tolerancing in the design, aimed at low sensitivity to drift.

Where adjustment is by a screwdriver or other tool, care should be taken to ensure that damage cannot be done to the equipment. Guide holes, for example, can prevent a screwdriver from slipping.

Where adjustment requires that measurements are made, or indicators observed, then the displays or meters should be easily visible while the adjustment is made.

It is usually necessary for adjustments and alignments to be carried out in a sequence and this must be specified in the maintenance instructions. The designer should understand that where drift in a particular component can be compensated for by the adjustment of some other item then, if that adjustment is difficult or critical, the service engineer will often change the drifting item, regardless of its cost.

14.1.3 Built-in test equipment

As with any test equipment, built-in test equipment (BITE) should be an order of magnitude more reliable than the system of which it is part, in order to minimize the incidence of false alarms or incorrect diagnosis. Poor-reliability BITE will probably reduce the system availability.

The number of connections between the system and the built-in test equipment should be minimized to reduce the probability of system faults induced by the BITE. It carries the disadvantages of being costly, inflexible (designed around the system it is difficult to modify) and of requiring some means of self-checking. In addition, it carries a weight, volume and power supply penalty but, on the other hand, greatly reduces the time required for realization diagnosis and checkout.

14.1.4 Circuit layout and hardware partitioning

It is advisable to consider maintainability when designing and laying out circuitry. In some cases it is possible to identify a logical sequence of events or signal flow through a circuit, and fault diagnosis is helped by a component layout which reflects this logic. Components should not be so close together as to make damage likely when removing and replacing a faulty item.

The use of integrated circuits introduces difficulties. Their small size and large number of leads make it necessary for connections to be small and close together, which increases the possibility of damage during maintenance. In any case, field maintenance at circuit level is almost impossible owing to the high function density involved. Because of the high maintenance cost of removing and resoldering these devices, the question of plug-in ICs arises. Another point of view emphasizes that IC sockets increase both cost and the possibility of connector failure. The decision for or against is made on economic grounds and must be taken on the basis of field failure rate, socket cost and repair time. The IC is a functional unit in itself and therefore circuit layout is less capable of representing the circuit function.

In general, the cost of microelectronics hardware continues to fall and thus the printed circuit board is more and more considered as a throwaway unit.

14.1.5 Connections

Connections present a classic trade-off between reliability and maintainability. The following types of connection are ranked in order of reliability, starting with the most reliable. A comparison of failure rates is made by means of the following:

Wrapped joint	0.00003 per 10^6 h
Welded connection	0.002 per 10^6 h
Machine-soldered joint	0.0003 per 10^6 h
Crimped joint	0.0003 per 10^6 h
Hand-soldered joint	0.0002 per 10^6 h
Edge connector (per pin)	0.001 per 10^6 h

Since edge connectors are less reliable than soldered joints, there needs to be a balance between having a few large plug-in units and a larger number of smaller throw-away units with the associated reliability problem of additional edge connectors. Boards terminated with wrapped joints rather than with edge connectors are two orders more reliable from the point of view of the connections, but the maintainability penalty can easily outweigh the reliability advantage. Bear in mind the time taken to make ten or twenty wrapped joints compared with that taken to plug in a board equipped with edge connectors.

The following are approximate times for making the different types of connection assuming that appropriate tools are available:

Edge connector (multi-contact)	10 s
Solder joint (single-wire)	20 s
Wrapped joint	50 s

As can be seen, maintainability ranks in the opposite order to reliability. In general, a high-reliability connection is required within the LRA, where maintainability is a secondary consideration. The interface between the LRA and the system requires a high degree of maintainability and the plug-in or edge connector is justified. If the LRA is highly reliable, and therefore unlikely to require frequent replacement, termination by the reliable wrapped joints could be justified. On the other hand a medium- or low-reliability unit would require plug and socket connection for quick interchange.

The reliability of a solder joint, hand or flow, is extremely sensitive to the quality control of the manufacturing process. Where cable connectors are used it should be ensured, by labelling or polarizing, that plugs will not be wrongly inserted in sockets or inserted in wrong sockets. Mechanical design should prevent insertion of plugs in the wrong configuration and also prevent damage to pins by clumsy insertion.

Where several connections are to be made within or between units, the complex of wiring is often provided by means of a cableform (loom) and the terminations (plug, solder or wrap) made according to an appropriate document. The cableform should be regarded as an LRA and local repairs should not be attempted. A faulty wire may be cut back, but left in place, and a single wire added to replace the link, provided that this does not involve the possibility of electrical pickup or misphasing.

14.1.6 Displays and indicators

Displays and indicators are effective in reducing the diagnostic, checkout and alignment contributions to active repair time. Simplicity should be the keynote and a 'go, no go' type of meter or display will require only a glance. The use of stark colour changes, or other obvious means, to divide a scale into areas of 'satisfactory operation' and 'alarm' should be used. Sometimes a meter, together with a multiway switch, is used to monitor several parameters in a system. It is desirable that the anticipated (normal) indication be the same for all the applications of the meter so that the correct condition is shown by little or no movement as the instrument is switched to the various test points. Displays should never be positioned where it is difficult, dangerous or uncomfortable to read them.

For an alarm condition an audible signal, as well as visual displays, is needed to draw attention to the fault. Displays in general, and those relating to alarm conditions in particular, must be more reliable than the parent system since a failure to indicate an alarm condition is potentially dangerous.

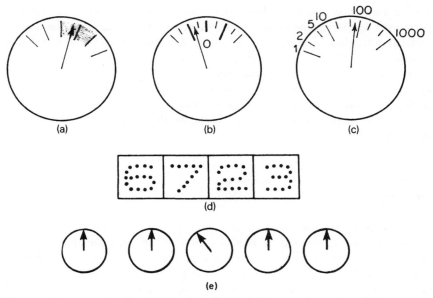

Figure 14.1 Meter displays. (a) Scale with shaded range; (b) scale with limits; (c) logarithmic scale; (d) digital display; (e) alignment of norms

If equipment is unattended then some alarms and displays may have to be extended to another location and the reliability of the communications link then becomes important to the availability of the system.

The following points concerning meters are worth noting:

1. False readings can result from parallax effects owing to scale and pointer being in different planes. A mirror behind the pointer helps to overcome this difficulty.
2. Where a range exists outside which some parameter is unacceptable, then either the acceptable or the unacceptable range should be coloured or otherwise made readily distinguishable from the rest of the scale (Figure 14.1(a)).
3. Where a meter displays a parameter which should normally have a single value, then a centre-zero instrument can be used to advantage and the circuitry configured such that the normal acceptable range of values falls within the mid-zone of the scale (Figure 14.1(b)).
4. Linear scales are easier to read and less ambiguous than logarithmic scales, and consistency in the choice of scales and ranges minimizes the possibility of misreading (Figure 14.1(c)). On the other hand, there are occasions when the use of a non-linear response or false-zero meter is desirable.
5. Digital displays are now widely used and are superior to the analogue pointer-type of instrument where a reading has to be recorded (Figure 14.1(d)). The analogue type of display is preferable when a check or adjustment within a range is required.
6. When a number of meters are grouped together it is desirable that the pointer positions for the *normal* condition are alike. Figure 14.1(e) shows how easily an incorrect reading is noticed.

Consistency in the use of colour codes, symbols and labels associated with displays is highly desirable. Filament lamps are not particularly reliable and should be derated. More reliable LEDs and liquid crystal displays are now widely used.

All displays should be positioned as near as possible to the location of the function or parameter to which they refer and mounted in an order relating to the sequence of adjustment. Unnecessary displays merely complicate the maintenance task and do more harm than good. Meters need be no more accurate than the measurement requirement of the parameter involved.

14.1.7 Handling, human and ergonomic factors

Major handling points to watch are:

● Weight, size and shape of removable modules. The LRA should not be capable of self-damage owing to its own instability, as in the case of a thin lamina construction.
● Protection of sharp edges and high-voltage sources. Even an unplugged module may hold dangerous charges on capacitors.
● Correct handles and grips reduce the temptation to use components for the purpose.
● When an inductive circuit is broken by the removal of a unit, then the earth return should not be via the frame. A separate earth return via a pin or connection from the unit should be used.

The following ergonomic factors also influence active repair time:

● Design for minimum maintenance skills considering what type of personnel are actually available.
● Beware of over-miniaturization - incidental damage is more likely.
● Consider comfort and safety of personnel when designing for access; e.g. body position, movements, limits of reach and span, limit of strength in various positions, etc.
● Illumination - fixed and portable.
● Shield from environment (weather, damp, etc.) and from stresses generated by the equipment (heat, vibration, noise, gases, moving parts, etc.) since repair is slowed down if the maintenance engineer has to combat these factors.

14.1.8 Identification

Identification of components, test points, terminals, leads, connectors and modules is helped by standardization of appearance. Colour codes should not be complex since over 5% of the male population suffer from some form of colour blindness. Simple, unambiguous, numbers and symbols help in the identification of particular functional modules. The physical grouping of functions simplifies the signs required to identify a particular circuit or LRA.

In many cases programmable hardware devices contain software (code). It is important to be able to identify the version of code resident in the device and this is often only possible by way of the component labelling.

14.1.9 Interchangeability

Where LRAs (Least Replaceable Assemblies, see section 14.1.10) are interchangeable this simplifies diagnosis, replacement and checkout, owing to the element of standardization involved. Spares provisioning then becomes slightly less critical in view of the possibility of using a non-essential, redundant, unit to effect a repair in some other part of the system. Cannibalization of several failed LRAs to yield a working module also becomes possible although this should never become standard field practice.

The smaller and less complex the LRA, the greater the possibility of standardization and hence interchangeability. The penalty lies in the number of interconnections, between LRAs and the system (less reliability) and the fact that the diagnosis is referred to a lower level (greater skill and more equipment).

Interchange of non-identical boards or units should be made mechanically impossible. At least, pin conventions should be such that insertion of an incorrect board cannot cause damage either to that board or to other parts of the equipment. Each value of power supply must always occupy the same pin number.

14.1.10 Least Replaceable Assembly

The LRA is that replaceable module at which local fault diagnosis ceases and direct replacement occurs. Failures are traced only to the LRA, which should be easily removable (see Section 14.1.5), replacement LRAs being the spares holding. It should rarely be necessary to remove an LRA in order to prove that it is faulty, and no LRA should require the removal of any other LRA for diagnosis or for replacement.

The choice of level of the LRA is one of the most powerful factors in determining maintainability. The larger the LRA, the faster the diagnosis. Maintainability, however, is not the only factor in the choice of LRA. As the size of the LRA increases so does its cost and the cost of spares holding. The more expensive the LRA, the less likely is a throw-away policy to be applicable. Also, a larger LRA is less likely to be interchangeable with any other. The following compares various factors as the size of LRA increases:

System maintainability	Improves
LRA reliability	Decreases
Cost of system testing (equipment and manpower)	Decreases
Cost of individual spares	Increases
Number of types of spares	Decreases

14.1.11 Mounting

If components are mounted so as to be self-locating then replacement is made easier. Mechanical design and layout of mounting pins and brackets can be made to prevent transposition where this is undesirable as in the case of a transformer, which must not be connected the wrong way round. Fragile components should be mounted as far as possible from handles and grips.

14.1.12 Component part selection

Main factors affecting repair times are:

Availability of spares – delivery.
Reliability/deterioration under storage conditions.
Ease of recognition.
Ease of handling.
Cost of parts.
Physical strength and ease of adjustment.

14.1.13 Redundancy

Circuit redundancy within the LRA (usually unmonitored) increases the reliability of the module, and this technique can be used in order to make it sufficiently reliable to be regarded as a throw-away unit. Redundancy at the LRA level permits redundant units to be removed for preventive maintenance while the system remains in service.

Although improving both reliability and maintainability, redundant units require more space and weight. Capital cost is increased and the additional units need more spares and generate more maintenance. System availability is thus improved but both preventive and corrective maintenance costs increase with the number of units.

14.1.14 Safety

Apart from legal and ethical considerations, safety-related hazards increase active repair time by requiring greater care and attention. An unsafe design will encourage short cuts or the omission of essential activities. Accidents add, very substantially, to the repair time.

Where redundancy exists, routine maintenance can be carried out after isolation of the unit from high voltage and other hazards. In some cases routine maintenance is performed under power, in which case appropriate safeguards must be incorporated into the design. The following practices should be the norm:

- Isolate high voltages under the control of microswitches which are automatically operated during access. The use of a positive interlock should bar access unless the condition is safe.
- Weights should not have to be lifted or supported.
- Use appropriate handles.
- Provide physical shielding from high voltage, high temperature, etc.
- Eliminate sharp points and edges.
- Install alarm arrangements. The exposure of a distinguishing colour when safety covers have been removed is good practice.
- Ensure adequate lighting.

14.1.15 Software

The availability of programmable LSI (large-scale integration) devices has revolutionized the approach to circuit design. More and more electronic circuitry is being replaced by a standard microprocessor architecture with the individual circuit requirements achieved within the software (program) which is held in the memory section of the hardware. Under these conditions diagnosis can no longer be supported by circuit descriptions and measurement information. Complex sequences of digital processing make diagnosis impossible with traditional test equipment.

Production testing of this type of printed-board assembly is possible only with sophisticated computer-driven automatic test equipment (ATE) and, as a result, field diagnosis can be only to board level. Where printed boards are interconnected by data highways carrying dynamic digital information, even this level of fault isolation may require field test equipment consisting of a microprocessor loaded with appropriate software for the unit under repair.

14.1.16 Standardization

Standardization leads to improved familiarization and hence shorter repair times. The number of different tools and test equipment is reduced, as is the possibility of delay due to having incorrect test gear. Fewer types of spares are required, reducing the probability of exhausting the stock.

14.1.17 Test points

Test points are the interface between test equipment and the system and are needed for diagnosis, adjustment, checkout, calibration and monitoring for drift. Their provision is largely governed by the level of LRA chosen and they will usually not extend beyond what is necessary to establish that an LRA is faulty. Test points within the LRA will be dictated by the type of board test carried out in production or in second-line repair.

In order to minimize faults caused during maintenance, test points should be accessible without the removal of covers and should be electrically buffered to protect the system from misuse of test equipment. Standard positioning also reduces the probability of incorrect diagnosis resulting from wrong connections. Test points should be grouped in such a way as to facilitate sequential checks. The total number should be kept to a minimum consistent with the diagnosis requirements. Unnecessary test points are likely to reduce rather than increase maintainability.

The above 17 design parameters relate to the equipment itself and not to the maintenance philosophy. Their main influence is on the active repair elements such as diagnosis, replacement, checkout, access and alignment. Maintenance philosophy and design are, nevertheless, interdependent. Most of the foregoing have some influence on the choice of test equipment. Skill requirements are influenced by the choice of LRA, by displays and by standardization. Maintenance procedures are affected by the size of modules and the number of types of spares. The following section will examine the ways in which maintenance philosophy and design act together to influence down times.

14.2 MAINTENANCE STRATEGIES AND HANDBOOKS

Both active and passive repair times are influenced by factors other than equipment design. Consideration of maintenance procedures, personnel, and spares provisioning is known as Maintenance Philosophy and plays an important part in determining overall availability. The costs involved in these activities are considerable and it is therefore important to strike a balance between over- and under-emphasizing each factor. They can be grouped under seven headings:

Organization of maintenance resources.
Maintenance procedures.
Tools and test equipment.
Personnel - selection, training and motivation.
Maintenance instructions and manuals.
Spares provisioning.
Logistics.

14.2.1 Organization of maintenance resources

It is usual to divide the maintenance tasks into three groups in order first, to concentrate the higher skills and more important test equipment in one place and second, to provide optimum replacement times in the field. These groups, which are known by a variety of names, are as follows.

First-line Maintenance – Corrective Maintenance – Call – Field Maintenance
This will entail diagnosis only to the level of the LRA, and repair is by LRA replacement. The technician either carries spare LRAs or has rapid access to them. Diagnosis may be aided by a portable intelligent terminal, especially in the case of microprocessor-based equipment. This group may involve two grades of technician, the first answering calls and the second being a small group of specialists who can provide backup in the more difficult cases.

Preventive Maintenance – Routine Maintenance
This will entail scheduled replacement/discard (see Chapter 16) of defined modules and some degree of cleaning and adjustment. Parametric checks to locate dormant faults and drift conditions may be included.

Second-line Maintenance – Workshop – Overhaul Shop – Repair Depot
This is for the purpose of:

1. Scheduled overhaul and refurbishing of units returned from preventive maintenance;
2. Unscheduled repair and/or overhaul of modules which have failed or become degraded.

Deeper diagnostic capability is needed and therefore the larger, more complex, test equipment will be found at the workshop together with full system information.

14.2.2 Maintenance procedures

For any of the above groups of staff it has been shown that fast, effective and error-free maintenance is best achieved if a logical and formal procedure is followed on each occasion. A haphazard approach based on the subjective opinion of the maintenance technician, although occasionally resulting in spectacular short cuts, is unlikely to prove the better method in the long run. A formal procedure also ensures that calibration and essential checks are not omitted, that diagnosis always follows a logical sequence designed to prevent incorrect or incomplete fault detection, that correct test equipment is used for each task (damage is likely if incorrect test gear is used) and that dangerous practices are avoided. Correct maintenance procedure is assured only by accurate and complete manuals and thorough training. A maintenance procedure must consist of the following:

Making and interpreting test readings;
Isolating the cause of a fault;
Part (LRA) replacement;
Adjusting for optimum performance (where applicable).

The extent of the diagnosis is determined by the level of fault identification and hence by the Least Replaceable Assembly. A number of procedures are used:

1. Stimuli - response: where the response to changes of one or more parameters is observed and compared with the expected response;
2. Parametric checks where parameters are observed at displays and test points and are compared with expected values;
3. Signal injection where a given pulse, or frequency, is applied to a particular point in the system and the signal observed at various points, in order to detect where it is lost, or incorrectly processed;
4. Functional isolation wherein signals and parameters are checked at various points, in a sequence designed to eliminate the existence of faults before or after each point. In this way, the location of the fault is narrowed down;
5. Robot test methods where automatic test equipment is used to fully 'flood' the unit with a simulated load, in order to allow the fault to be observed.

Having isolated the fault, a number of repair methods present themselves:

1. Direct replacement of the LRA;
2. Component replacement or rebuilding, using simple construction techniques;
3. Cannibalization from non-essential parts.

In practice, direct replacement of the LRA is the usual solution owing to the high cost of field repair and the need for short down times in order to achieve the required equipment availability.

Depending upon circumstances, and the location of a system, repair may be carried out either immediately a fault is signalled or only at defined times, with redundancy being relied upon to maintain service between visits. In the former case, system reliability depends on the mean repair time and in the latter, upon the interval between visits and the amount of redundancy provided.

14.2.3 Tools and test equipment

The following are the main considerations when specifying tools and test equipment.

1. Simplicity: test gear should be easy to use and require no elaborate set-up procedure.
2. Standardization: the minimum number of types of test gear reduces the training and skill requirements and minimizes test equipment spares holdings. Standardization should include types of displays and connections.
3. Reliability: test gear should be an order of magnitude more reliable than the system for which it is designed, since a test equipment failure can extend down time or even result in a system failure.
4. Maintainability: ease of repair and calibration will affect the non-availability of test gear. Ultimately it reduces the amount of duplicate equipment required.
5. Replacement: suppliers should be chosen bearing in mind the delivery time for replacements and for how many years they will be available.

There is a trade-off between the complexity of test equipment and the skill and training of maintenance personnel. This extends to built-in test equipment (BITE) which, although introducing some disadvantages, speeds and simplifies maintenance.

BITE forms an integral part of the system and requires no setting-up procedure in order to initiate a test. Since it is part of the system, weight, volume and power consumption are important. A customer may specify these constraints in the system specification (e.g. power requirements of BITE not to exceed 2% of mean power consumption). Simple BITE can be in the form of displays of various parameters. At the other end of the scale, it may consist of a programmed sequence of stimuli and tests, which culminate in a 'print-out' of diagnosis and repair instructions. There is no simple formula, however, for determining the optimum combination of equipment complexity and human skill. The whole situation, with the variables mentioned, has to be considered and a trade-off technique found which takes account of the design parameters together with the maintenance philosophy.

There is also the possibility of Automatic Test Equipment (ATE) being used for field maintenance. In this case, the test equipment is quite separate from the system and is capable of monitoring several parameters simultaneously and on a repetitive basis. Control is generally by software and the maintenance task is simplified.

When choosing simple portable test gear, there is a choice of commercially available general-purpose equipment, as against specially designed equipment. Cost and ease of replacement favour the general-purpose equipment whereas special-purpose equipment can be made simpler to use and more directly compatible with test points.

In general, the choice between the various test equipment options involves a trade-off of complexity, weight, cost, skill levels, time scales and design, all of which involve cost, with the advantages of faster and simpler maintenance.

14.2.4 Personnel considerations

Four staffing considerations influence the maintainability of equipment:

Training given
Skill level employed
Motivation
Quantity and distribution of personnel

More complex designs involve a wider range of maintenance and hence more training is required. Proficiency in carrying out corrective maintenance is achieved by a combination of knowledge and diagnostic skill. Whereas knowledge can be acquired by direct teaching methods, skill can be gained only from experience, in either a simulated or a real environment. Training must, therefore, include experience of practical fault finding on actual equipment. Sufficient theory, in order to understand the reasons for certain actions and to permit logical reasoning, is required, but an excess of theoretical teaching is both unnecessary and confusing. A balance must be achieved between the confusion of too much theory and the motivating interest created by such knowledge.

A problem with very high-reliability equipment is that some failure modes occur so infrequently that the technicians have little or no field experience of their diagnosis and repair. Refresher training with simulated faults will be essential to ensure effective maintenance, should it be required. Training maintenance staff in a variety of skills (e.g. electronic as well as electromechanical work) provides a flexible workforce and reduces the probability of a technician being unable to deal with a particular failure unaided. Less time is wasted during a repair and transport costs are also reduced.

Training of customer maintenance staff is often given by the contractor, in which case an objective test of staff suitability may be required. Well-structured training which provides flexibility and proficiency, improves motivation since confidence, and the ability to perform a number of tasks, brings job satisfaction in demonstrating both speed and accuracy. In order to achieve a given performance, specified training and a stated level of ability are assumed. Skill levels must be described in objective terms of knowledge, dexterity, memory, visual acuity, physical strength, inductive reasoning and so on.

Staff scheduling requires a knowledge of the equipment failure rates. Different failure modes require different repair times and have different failure rates.

The MTTR may be reduced by increasing the effort from one to two technicians but any further increase in personnel may be counter-productive and not significantly reduce the repair time.

Personnel policies are usually under the control of the customer and, therefore, close liaison between contractor and customer is essential before design features relating to maintenance skills can be finalized. In other words, the design specification must reflect the personnel aspects of the maintenance philosophy.

14.2.5 Maintenance manuals

Requirements

The main objective of a maintenance manual is to provide all the information required to carry out each maintenance task without reference to the base workshop, design authority or any other source of information. It may, therefore, include any of the following:

- Specification of system performance and functions.
- Theory of operation and usage limitations.
- Method of operation.
- Range of operating conditions.
- Supply requirements.
- Corrective and preventive maintenance routines.
- Permitted modifications.
- Description of spares and alternatives.
- List of test equipment and its check procedure.
- Disposal instructions for hazardous materials.

The actual manual might range from a simple card, which could hang on a wall, to a small library of information comprising many handbooks for different applications and users. Field reliability and maintainability are influenced, in no small way, by the maintenance instructions. The design team, or the maintainability engineer, has to supply information to the handbook writer and to collaborate if the instructions are to be effective.

Consider the provision of maintenance information for a complex system operated by a well-managed organization. The system will be maintained by a permanent team (A) based on site. This team of technicians, at a fair level of competence, service a range of systems and, therefore, are not expert in any one particular type of equipment. Assume that the system incorporates some internal monitoring equipment and that specialized portable test gear is available for both fault diagnosis and for routine checks. This local team carries out all the routine checks and repairs most faults by means of module replacement. There is a limited local stock of some modules (LRAs) which is replenished from a central depot which serves several sites. The depot also stocks those replacement items not normally held on-site.

Based at the central depot is a small staff of highly skilled specialist technicians (B) who are available to the individual sites. Available to them is further specialized test gear and also basic instruments capable of the full range of measurements and tests likely to be made. These technicians are called upon when the first-line (on-site) procedures are inadequate for diagnosis or replacement. This team also visits the sites in order to carry out the more complex or critical periodic checks.

Also at the central depot is a workshop staffed with a team of craftsmen and technicians (C) who carry out the routine repairs and the checkout of modules returned from the field. The specialist team (B) is available for diagnosis and checkout whenever the (C) group is unable to repair modules.

A maintenance planning group (D) is responsible for the management of the total service operation, including cost control, coordination of reliability and maintainability statistics, system modifications, service manual updating, spares provisioning, stock control and, in some cases, a post-design service.

A preventive maintenance team (E), also based at the depot, carries out the regular replacements and adjustments to a strict schedule.

Group A will require detailed and precise instructions for the corrective tasks which it carries out. A brief description of overall system operation is desirable to the extent of stimulating interest but it should not be so detailed as to permit unorthodox departures from the maintenance instructions. There is little scope for initiative in this type of maintenance since speedy module diagnosis and replacement is required. Instructions for incident reporting should be included and a set format used.

Group B requires a more detailed set of data since it has to carry out fault diagnosis in the presence of intermittent, marginal or multiple faults not necessarily anticipated when the handbooks were prepared. Diagnosis should nevertheless still be to LRA level since the philosophy of first-line replacement holds.

Group C will require information similar to that of Group A but will be concerned with the diagnosis and repair of modules. It may well be that certain repairs require the fabrication of piece parts, in which case the drawings and process instructions must be available.

Group D requires considerable design detail and a record of all changes. This will be essential after some years of service when the original design team may not be available to give advice. Detailed spares requirements are essential so that adequate, safe substitutions can be made in the event of a spares source or component type becoming unavailable. Consider a large population item which may have been originally subject to stringent screening for high reliability. Obtaining a further supply in a small quantity but to the same standard may be impossible, and their replacement with less-assured items may have to be considered. Consider also an item selected to meet a wide range of climatic conditions. A particular user may well select a cheaper replacement meeting his or her own conditions of environment.

Group E requires detailed instructions since, again, little initiative is required. Any departure from the instructions implies a need for Group A.

Types of manual
Preventive maintenance procedures will be listed in groups by service intervals, which can be by calendar time, switch-on time, hours flown, miles travelled, etc., as appropriate. As with calibration intervals, the results and measurements at each maintenance should be used to lengthen or shorten the service interval as necessary. The maintenance procedure and reporting

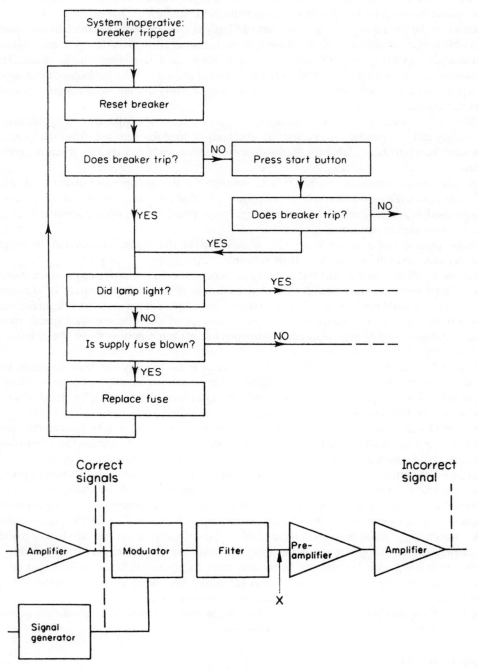

Figure 14.2

requirements must be very fully described so that little scope for initiative or interpretation is required. In general, all field maintenance should be as routine as possible and capable of being fully described in a manual. Any complicated diagnosis should be carried out at the workshop and module replacement on-site used to achieve this end. In the event of a routine maintenance check not yielding the desired result, the technician should either be referred to the corrective maintenance procedure or told to replace the suspect module.

In the case of *corrective maintenance* (callout for failure or incident) the documentation should first list all the possible indications such as print-outs, alarms, displays, etc. Following this, routine functional checks and test point measurements can be specified. This may involve the use of a portable 'intelligent' terminal capable of injecting signals and making decisions based on the responses. A fault dictionary is a useful aid and should be continuously updated with data from the field and/or design and production areas. Full instructions should be included for isolating parts of the equipment or taking precautions where safety is involved. Precautions to prevent secondary failures being generated should be thought out by the designer and included in the maintenance procedure.

Having isolated the fault and taken any necessary precautions, the next consideration is the diagnostic procedure followed by repair and checkout. Diagnostic procedures are best described in a logical flow chart. Figure 14.2 shows a segment of a typical diagnostic algorithm involving simple Yes/No decisions with paths of action for each branch. Where such a simple process is not relevant and the technician has to use initiative, then the presentation of schematic diagrams and the system and circuit descriptions are important. Some faults, by their nature or symptoms, indicate the function which is faulty and the algorithm approach is most suitable. Other faults are best detected by observing the conditions existing at the interfaces between physical assemblies or functional stages. Here the location of the fault may be by a bracketing/ elimination process. For example 'The required signal appears at point 12 but is not present at point 20. Does it appear at point 16? No, but it appears at point 14. Investigate unit between points 14 and 16'. The second part of Figure 14.2 is an example of this type of diagnosis presented in a flow diagram. In many cases a combination of the two approaches may be necessary.

14.2.6 Spares provisioning

Figure 14.3 shows a simple model for a system having n of a particular item and a nominal spares stock of r. The stock is continually replenished either by repairing failed items or by ordering new spares. In either case the repair time or lead time is shown as T. It is assumed that the system repair is instantaneous, given that a spare is available. Then the probability of a stockout causing system failure is given by a simple statistical model. Let the system

Figure 14.3 Spares replacement from second-line repair

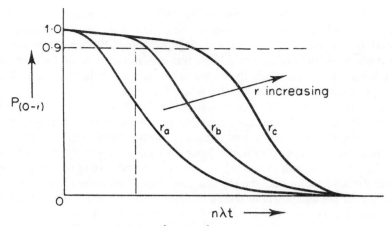

Figure 14.4 Set of curves for spares provisioning

λ = failure rate

n = number of parts of that type which may have to be replaced

r = number of spares of that part carried

$P_{(0-r)}$ = probability of 0 - r failures = probability of stock not being exhausted

Unavailability be U and assume that failures occur at random allowing a constant failure rate model to be used.

U = 1 – Probability of stock not being exhausted

= 1 – Probability of 0 to r failures in T.

Figure 14.4 shows a set of simple Poisson curves which give P_{0-r} against $n\lambda T$ for various values of spares stock, r. The curves in Chapter 5 are identical and may be used to obtain answers based on this model.

A more realistic, and therefore more complex, Unavailability model would take account of two additional parameters:

- The Down Time of the system while the spare (if available) is brought into use and the repair carried out;
- Any redundancy. The simple model assumed that all n items were needed to operate. If some lesser number were adequate then a partial redundancy situation would apply and the Unavailability would be less.

The simple Poisson model will not suffice for this situation and a more sophisticated technique, namely the Markov method described in Chapter 8, is needed for the calculations.

Figure 14.5 shows a typical state diagram for a situation involving 4 units and 2 spares. The lower left hand state represents 4 good items, with none failed and 2 spares. This is the 'start' state. A failure (having the rate 4λ) brings the system to the state, immediately to the right, where there are 3 operating with one failure but still 2 spares. The transition diagonally upwards to the left represents a repair (i.e. replacement by a spare). The subsequent transition downwards represents a procurement of a new spare and brings the system back to the 'start' state. The other states and transitions model the various possibilities of failure and spares states for the system.

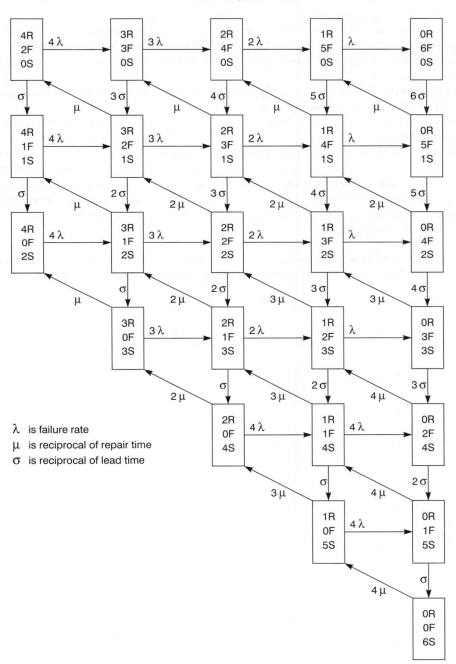

Figure 14.5 Markov state diagram

If no redundancy exists then the availability (1-unavailability) is obtained by evaluating the probability of being in any of the 3 states shown in the left hand column of the state diagram. '3 out of 4' redundancy would imply that the availability is obtained from considering the probability of being in any of the states in the first 2 left hand columns, and so on.

Numerical evaluation of these states is obtained from the computer package COMPARE for each case of Number of Items, Procurement Time and Repair Time. Values of unavailability can be obtained for a number of failure rates and curves are then drawn for each case to be assessed.

The appropriate failure rate for each item can then be used to assess the Unavailability associated with each of various spares levels.

Figure 14.6 gives an example of Unavailability curves for specific values of MDT, turnround time and redundancy.

The curves show the Unavailability against failure rate for 0, 1, and 2 spares. The curve for infinite spares gives the Unavailability based only on the 12 hours Down Time. It can only be seen in Figure 14.6 by understanding that for all values greater than 2 spares the line cannot be distinguished from the 2+ line. In other words, for 2 spares and greater, the Unavailability is dominated by the repair time. For that particular example the following observations might be made when planning spares:

- For failure rates greater than about 25×10^{-6} per hour the Unavailability is still significant even with large numbers of spares. Attention should be given to reducing the down time.

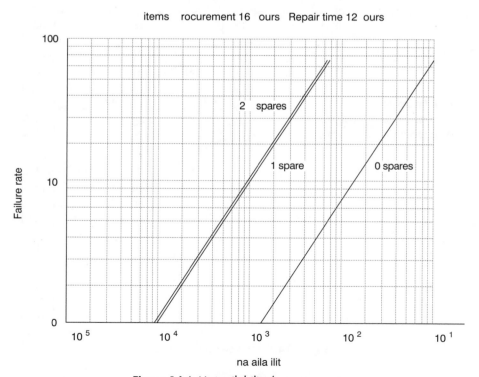

Figure 14.6 Unavailability/spares curves

- For failure rates less than about 3×10^{-6} per hour one spare is probably adequate and no further analysis is required.

It must be stressed that this is only one specific example and that the values will change considerably as the different parameters are altered.

The question arises as to whether spares that have been repaired should be returned to a central stock or retain their identity for return to the parent system. Returning a part to its original position is costly and requires a procedure so that initial replacement is only temporary. This may be necessary where servicing is carried out on equipment belonging to different customers – indeed some countries impose a legal requirement to this end. Another reason for retaining a separate identity for each unit occurs owing to wearout, when it is necessary to know the expired life of each item.

Stock control is necessary when holding spares and inputs are therefore required from:

Preventive and corrective maintenance in the field
Second-line maintenance
Warranty items supplied

The main considerations of spares provisioning are:

1. Failure rate – determines quantity and perhaps location of spares.
2. Acceptable probability of stockout – fixes spares level.
3. Turnround of second-line repair – affects lead time.
4. Cost of each spare – affects spares level and hence item 2.
5. Standardization and LRA – affects number of different spares to be held.
6. Lead time on ordering – effectively part of second-line repair time.

14.2.7 Logistics

Logistics is concerned with the time and resources involved in transporting personnel, spares and equipment into the field. The main consideration is the degree of centralization of these resources.

Centralize	*Decentralize*
Specialized test equipment.	Small tools and standard items.
Low utilization of skills and test gear.	Where small MTTR is vital.
Second-line repair.	Fragile test gear.
Infrequent (high-reliability) spares.	Frequent (low-reliability) spares.

A combination will be found where a minimum of on-site facilities, which ensures repair within the specified MTTR, is provided. The remainder of the spares backup and low utilization test gear can then be centralized. If Availability is to be kept high by means of a low MTTR then spares depots have to be established at a sufficient number of points to permit access to spares within a specified time.

14.2.8 The user and the designer

The considerations discussed in this chapter are very much the user's concern. It is necessary, however, to decide upon them at the design stage since they influence, and are influenced by, the engineering of the product. The following table shows a few of the relationships between maintenance philosophy and design.

Skill level of maintenance technician	Amount of built-in test equipment required Level of LRA replacement in the field
Tools and test equipment	LRA fixings, connections and access Test points and equipment standardization
Maintenance procedure	Ergonomics and environment Built-in test equipment diagnostics Displays Interchangeability

The importance of user involvement at the very earliest stages of design cannot be over-emphasized. Maintainability objectives cannot be satisfied merely by placing requirements on the designer and neither can they be considered without recognizing that there is a strong link between repair time and cost. The maintenance philosophy has therefore to be agreed while the design specification is being prepared.

14.2.9 Computer aids to maintenance

The availability of computer packages makes it possible to set up a complete preventive maintenance and spare-part provisioning scheme using computer facilities. The system is described to the computer by delineating all the parts and their respective failure rates, and routine maintenance schedules and the times to replenish each spare. The operator will then receive daily schedules of maintenance tasks with a list of spares and consumables required for each. There is automatic indication when stocks of any particular spare fall below the minimum level.

These minimum spares levels can be calculated from a knowledge of the part failure rate and ordering time if a given risk of spares stockout is specified.

Packages exist for optimum maintenance times and spares levels. The COMPARE package offers the type of Reliability Centred Maintenance calculations described in Chapter 16.

15 Predicting and demonstrating repair times

15.1 PREDICTION METHODS

The best-known methods for Maintainability prediction are described in US Military Handbook 472. The methods described in this Handbook, although applicable to a range of equipment developed at that time, have much to recommend them and are still worth attention. Unfortunately, the quantity of data required to develop these methods of prediction is so great that, with increasing costs and shorter design lives, it is unlikely that models will continue to be developed. On the other hand, calculations requiring the statistical analysis of large quantities of data lend themselves to computer methods and the rapid increase of these facilities makes such a calculation feasible if the necessary repair time data for a very large sample of repairs (say, 10 000) are available.

Any realistic maintainability prediction procedure must meet the following essential requirements:

1. The prediction must be fully documented and described and subject to recorded modification as a result of experience.
2. All assumptions must be recorded and their validity checked where possible.
3. The prediction must be carried out by engineers who are not part of the design group and therefore not biased by the objectives.

Prediction, valuable as it is, should be followed by demonstration as soon as possible in the design programme. Maintainability is related to reliability in that the frequency of each repair action is determined by failure rates. Maintainability prediction therefore requires a knowledge of failure rates in order to select the appropriate, weighted, sample of tasks. The prediction results can therefore be no more reliable than the accuracy of the failure rate data. Prediction is applicable only to the active elements of repair time since it is those which are influenced by the design.

There are two approaches to the prediction task. The FIRST is a work study method which analyses each task in the sample by breaking it into definable work elements. This requires an extensive data bank of average times for a wide range of tasks on the equipment type in question. The SECOND approach is empirical and involves rating a number of maintainability factors against a checklist. The resulting 'scores' are converted to an MTTR by means of a nomograph which was obtained by regression analysis of the data.

The methods (called procedures) in US Military Handbook 472 are over 20 years old and it is unlikely that the databases are totally relevant to modern equipment. In the absence of alternative methods, however, procedure 3 is recommended because the prediction will still give

Equip._____   Unit/Part_____   Task No._____

Ass'y_____   By_____   Date_____

Primary function failed unit/part_____

_____

Mode of failure_____

Malfunction symptoms_____

_____

Maintenance Analysis

Maintenance Steps	Scoring Comments

Checklist Scores

	1	2	3	4	5	6	7	8	9	10	11	12	13	14	15	Total
A																
B																
C																

Predicted downtime   ____ Min.

Figure 15.1

a fair indication of the repair time and also because the checklist approach focuses attention on the practical features affecting repair time. Procedure 3 is therefore described here in some detail.

15.1.1 US Military Handbook 472 – Procedure 3

Procedure 3 was developed by RCA for the US Air Force and was intended for ground systems. It requires a fair knowledge of the design detail and maintenance procedures for the system being analysed. The method is based on the principle of predicting a sample of the maintenance tasks. It is entirely empirical since it was developed to agree with known repair times for specific systems, including search radar, data processors and a digital data transmitter with r.f. elements. The sample of repair tasks is selected on the basis of failure rates and it is assumed

Figure 15.2

that the time to diagnose and correct a failure of a given component is the same as for any other of that component type. This is not always true, as field data can show.

Where repair of the system is achieved by replacement of sizeable modules (that is, a large LRA) the sample is based on the failure rate of these high-level units.

The predicted repair time for each sample task is arrived at by considering a checklist of maintainability features and by scoring points for each feature. The score for each feature increases with the degree of conformity with a stated 'ideal'. The items in the checklist are grouped under three headings: Design, Maintenance Support and Personnel Requirements. The points scored under each heading are appropriately weighted and related to the predicted repair time by means of a regression equation which is presented in the form of an easily used nomograph.

Figure 15.1 shows the score sheet for use with the checklist and Figure 15.2 presents the regression equation nomograph. I deduce the regression equation to be:

$$\log_{10}\text{MTTR} = 3.544 - 0.0123C - 0.023(1.0638A + 1.29B)$$

where A, B and C are the respective checklist scores.

Looking at the checklist it will be noted that additional weight is given to some features of design or maintenance support by the fact that more than one score is influenced by a particular feature.

The checklist is reproduced, in part, in the following section but the reader wishing to carry out a prediction will need a copy of US Military Handbook 472 for the full list. The application of the checklist to typical tasks is, in the author's opinion, justified as an aid to maintainability design even if repair time prediction is not specifically required.

15.1.2 Checklist – Mil 472 – Procedure 3

The headings of each of the checklists are as follows:

Checklist A

1. Access (external)
2. Latches and fasteners (external)
3. Latches and fasteners (internal)
4. Access (internal)
5. Packaging
6. Units/parts (failed)
7. Visual displays
8. Fault and operation indicators
9. Test points availability
10. Test points identification
11. Labelling
12. Adjustments
13. Testing in circuit
14. Protective devices
15. Safety – personnel

Checklist B

1. External test equipment
2. Connectors
3. Jigs and fixtures
4. Visual contact
5. Assistance operations
6. Assistance technical
7. Assistance supervisory

Checklist C

1. Arm – leg – back strength
2. Endurance and energy
3. Eye – hand
4. Visual
5. Logic
6. Memory
7. Planning
8. Precision
9. Patience
10. Initiative

Three items from each of checklists A and B and the scoring criteria for all of checklist C are reproduced as follows.

Checklist A – Scoring Physical Design Factors

(1) *Access (external)*: Determines if the external access is adequate for visual inspection and manipulative actions. Scoring will apply to external packaging as related to maintainability design concepts for ease of maintenance. This item is concerned with the design for external visual and manipulative actions which would precede internal maintenance actions. The following scores and scoring criteria will apply:

Scores

(a) Access adequate both for visual and manipulative tasks (electrical and mechanical)   4
(b) Access adequate for visual, but not manipulative, tasks   2
(c) Access adequate for manipulative, but not visual, tasks   2
(d) Access not adequate for visual or manipulative tasks   0

Scoring criteria

An explanation of the factors pertaining to the above scores is consecutively shown. This procedure is followed throughout for other scores and scoring criteria.

(a) To be scored when the external access, while visual and manipulative actions are being performed on the exterior of the subassembly, does not present difficulties because of obstructions (cables, panels, supports, etc.).
(b) To be scored when the external access is adequate (no delay) for visual inspection, but not for manipulative actions. External screws, covers, panels, etc., can be located visually; however, external packaging or obstructions hinders manipulative actions (removal, tightening, replacement, etc.).
(c) To be scored when the external access is adequate (no delay) for manipulative actions, but not for visual inspections. This applies to the removal of external covers, panels, screws, cables, etc., which present no difficulties; however, their location does not easily permit visual inspection.
(d) To be scored when the external access is inadequate for both visual and manipulative tasks. External covers, panels, screws, cables, etc., cannot be easily removed nor visually inspected because of external packaging or location.

(2) *Latches and fasteners (external)*: Determines if the screws, clips, latches, or fasteners outside the assembly require special tools, or if significant time was consumed in the removal of such items. Scoring will relate external equipment packaging and hardware to maintainability design concepts. Time consumed with preliminary external disassembly will be proportional to the type of hardware and tools needed to release them and will be evaluated accordingly.

Scores

(a) External latches and/or fasteners are captive, need no special tools, and require only a fraction of a turn for release   4
(b) External latches and/or fasteners meet two of the above three criteria   2
(c) External latches and/or fasteners meet one or none of the above three criteria   0

Scoring criteria

(a) To be scored when external screws, latches, and fasteners are:

   (1) Captive
   (2) Do not require special tools
   (3) Can be released with a fraction of a turn

   Releasing a 'DZUS' fastener which requires a 90-degree turn using a standard screwdriver is an example of all three conditions.
(b) To be scored when external screws, latches, and fasteners meet two of the three conditions stated in (a) above. An action requiring an Allen wrench and several full turns for release shall be considered as meeting only one of the above requirements.
(c) To be scored when external screws, latches, and fasteners meet only one or none of the three conditions stated in (a) above.

(3) *Latches and fasteners (internal)*: Determines if the internal screws, clips, fasteners or latches within the unit require special tools, or if significant time was consumed in the removal of such items. Scoring will relate internal equipment hardware to maintainability design concepts. The types of latches and fasteners in the equipment and standardization of these throughout the equipment will tend to affect the task by reducing or increasing required time to remove and replace them. Consider 'internal' latches and fasteners to be within the interior of the assembly.

Scores

(a) Internal latches and/or fasteners are captive, need no special tools, and require only a fraction of a turn for release   4
(b) Internal latches and/or fasteners meet two of the above three criteria   2
(c) Internal latches and/or fasteners meet one or none of the above three criteria   0

Scoring criteria

(a) To be scored when internal screws, latches and fasteners are:

   (1) Captive
   (2) Do not require special tools
   (3) Can be released with a fraction of a turn

   Releasing a 'DZUS' fastener which requires a 90-degree turn using a standard screwdriver would be an example of all three conditions.
(b) To be scored when internal screws, latches, and fasteners meet two of the three conditions stated in (a) above. A screw which is captive can be removed with a standard or Phillips screwdriver, but requires several full turns for release.
(c) To be scored when internal screws, latches, and fasteners meet one of three conditions stated in (a) above. An action requiring an Allen wrench and several full turns for release shall be considered as meeting only one of the above requirements.

Checklist B – Scoring design dictates – facilities

The intent of this questionnaire is to determine the need for external facilities. Facilities, as used here, include material such as test equipment, connectors, etc., and technical assistance from other maintenance personnel, supervisor, etc.

(1) *External test equipment*: Determines if external test equipment is required to complete the maintenance action. The type of repair considered maintainably ideal would be one which did not require the use of external test equipment. It follows, then, that a maintenance task requiring test equipment would involve more task time for set-up and adjustment and should receive a lower maintenance evaluation score.

Scores

(a) Task accomplishment does not require the use of external test equipment   4
(b) One piece of test equipment is needed   2
(c) Several pieces (2 or 3) of test equipment are needed   1
(d) Four or more items are required   0

Scoring criteria

(a) To be scored when the maintenance action does not require the use of external test equipment. Applicable when the cause of malfunction is easily detected by inspection or built-in test equipment.
(b) To be scored when one piece of test equipment is required to complete the maintenance action. Sufficient information is available through the use of one piece of external test equipment for adequate repair of the malfunction.
(c) To be scored when 2 or 3 pieces of external test equipment are required to complete the maintenance action. This type of malfunction would be complex enough to require testing in a number of areas with different test equipment.
(d) To be scored when four or more pieces of test equipment are required to complete the maintenance action. Involves an extensive testing requirement to locate the malfunction. This would indicate that a least maintainable condition exists.

(2) *Connectors*: Determines if supplementary test equipment requires special fittings, special tools, or adaptors to adequately perform tests on the electronic system or subsystem. During troubleshooting of electronic systems, the minimum need for test equipment adaptors or connectors indicates that a better maintainable condition exists.

Scores

(a) Connectors to test equipment require no special tools, fittings, or adaptors   4
(b) Connectors to test equipment require some special tools, fittings, or adaptors
    (less than two)   2
(c) Connectors to test equipment require special tools, fittings, and adaptors
    (more than one)   0

Scoring criteria

(a) To be scored when special fittings or adaptors and special tools are not required for testing. This would apply to tests requiring regular test leads (probes or alligator clips) which can be plugged into or otherwise secured to the test equipment binding post.

(b) Applies when one special fitting, adaptor or tool is required for testing. An example would be if testing had to be accomplished using a 10 dB attenuator pad in series with the test set.

(c) To be scored when more than one special fitting, adaptor, or tool is required for testing. An example would be when testing requires the use of an adaptor and an r.f. attenuator.

(3) *Jigs or fixtures*: Determines if supplementary materials such as block and tackle, braces, dollies, ladder, etc., are required to complete the maintenance action. The use of such items during maintenance would indicate the expenditure of a major maintenance time and pinpoint specific deficiencies in the design for maintainability.

Scores

(a) No supplementary materials are needed to perform task   4
(b) No more than one piece of supplementary material is needed to
perform task   2
(c) Two or more pieces of supplementary material are needed   0

Scoring criteria

(a) To be scored when no supplementary materials (block and tackle, braces, dollies, ladder, etc.) are required to complete maintenance. Applies when the maintenance action consists of normal testings and the removal or replacement of parts or components can be accomplished by hand, using standard tools.
(b) To be scored when one supplementary material is required to complete maintenance. Applies when testing or when the removal and replacement of parts requires a step ladder for access or a dolly for transportation.
(c) To be scored when more than one supplementary material is required to complete maintenance. Concerns the maintenance action requiring a step ladder and dolly adequately to test and remove the replaced parts.

Checklist C – Scoring design dictates – maintenance skills

This checklist evaluates the personnel requirements relating to physical, mental, and attitude characteristics, as imposed by the maintenance task.

Evaluation procedure for this checklist can best be explained by way of several examples. Consider the first question which deals with arm, leg and back strength. Should a particular task require the removal of an equipment drawer weighing 100 pounds, this would impose a severe requirement on this characteristic. Hence, in this case the question would be given a low score (0 to 1). Assume another task which, owing to small size and delicate construction, required extremely careful handling. Here question 1 would be given a high score (4), but the question dealing with eye-hand coordination and dexterity would be given a low score. Other questions in the checklist relate to various personnel characteristics important to maintenance task accomplishment. In completing the checklist, the task requirements that each of these characteristics should be viewed with respect to average technician capabilities.

Scores

	Score
1. Arm, leg, and back strength	_____
2. Endurance and energy	_____

Figure 15.3 MTTR demonstration test plan

3. Eye-hand coordination, manual dexterity, and neatness   _____
4. Visual acuity   _____
5. Logical analysis   _____
6. Memory – things and ideas   _____
7. Planfulness and resourcefulness   _____
8. Alertness, cautiousness, and accuracy   _____
9. Concentration, persistence and patience   _____
10. Initiative and incisiveness   _____

Scoring criteria

Quantitative evaluations of these items range from 0 to 4 and are defined in the following manner:

4. The maintenance action requires a minimum effort on the part of the technician.
3. The maintenance action requires a *below average* effort on the part of the technician.
2. The maintenance action requires an *average* effort on the part of the technician.
1. The maintenance action requires an *above average* effort on his part.
0. The maintenance action requires a *maximum* effort on his part.

15.2 DEMONSTRATION PLANS

15.2.1 Demonstration risks

Where demonstration of maintainability is contractual, it is essential that the test method, and the conditions under which it is to be carried out, are fully described. If this is not observed then disagreements are likely to arise during the demonstration. Both supplier and customer wish to achieve the specified Mean Time To Repair at minimum cost and yet a precise demonstration having acceptable risks to all parties is extremely expensive. A true assessment of maintainability can only be made at the end of the equipment life and anything less will represent a sample.

Figure 15.3 shows a typical test plan for observing the Mean Time To Repair of a given item. Just as, in Chapter 5, the curve shows the relationship of the probability of passing the test against the batch failure rate, then Figure 15.3 relates that probability to the actual MTTR.

Figure 15.4 Distribution of repair times

For a MTTR of M_0 the probability of passing the test is 90% and for a value of M_1 it falls to 10%. In other words, if M_0 and M_1 are within 2 : 1 of each other then the test has a good discrimination.

A fully documented procedure is essential and the only reference document available is US Military Standard 471A – Maintainability Verification/Demonstration/Evaluation – 27 March 1973. This document may be used as the basis for a contractual agreement in which case both parties should carefully assess the risks involved. Statistical methods are usually dependent on assumptions concerning the practical world and it is important to establish their relevance to a particular test situation. In any maintainability demonstration test it is absolutely essential to fix the following:

Method of test demonstration task selection
Tools and test equipment available
Maintenance documentation
Skill level and training of test subject
Environment during test
Preventive maintenance given to test system

15.2.2 US Mil Standard 471A (1973)

This document replaces US Military Standard 471 (1971) and MIL473 (1971) – Maintainability Demonstration. It contains a number of sampling plans for demonstrating maintenance times for various assumptions of repair time distribution. A task sampling plan is also included and describes how the sample of simulated failures should be chosen. Test plans choose either the log normal assumption or make no assumption of distribution. The log normal distribution frequently applies to systems using consistent technologies such as computer and data systems, telecommunications equipment, control systems and consumer electronics, but equipment with mixed technologies such as aircraft flight controls, microprocessor-controlled mechanical equipment and so on are likely to exhibit bimodal distributions. This results from two repair time distributions (for two basic types of defect) being superimposed. Figure 15.4 illustrates this case.

The method of task sample selection involves stratified sampling. This involves dividing the equipment into functional units and, by ascribing failure rates to each unit, determining the relative frequency of each maintenance action. Taking into account the quantity of each unit the sample of tasks is spread according to the anticipated distribution of field failures. Random sampling is used to select specific tasks within each unit once the appropriate number of tasks has been assigned to each. The seven test plans are described as follows:

Test Method 1

The method tests for the mean repair time (MTTR). A minimum sample size of 30 is required and an equation is given for computing its value. Equations for the producer's and consumer's risks, α and β, and their associated repair times are also given. Two test plans are given. Plan A assumes a log normal distribution of repair times while plan B is distribution free. That is, it applies in all cases.

Test Method 2

The method tests for a percentile repair time. This means a repair time associated with a given probability of not being exceeded. For example, a 90 percentile repair time of one hour means that 90% of repairs are effected in one hour or less and that only 10% exceed this value. This test assumes a log normal distribution of repair times. Equations are given for calculating the sample size, the risks and their associated repair times.

Test Method 3

The method tests the percentile value of a specified repair time. It is distribution free and therefore applies in all cases. For a given repair time, values of sample size and pass criterion are calculated for given risks and stated pass and fail percentiles. For example, if a median MTTR of 30 min is acceptable, and if 30 min as the 25th percentile (75% of values are greater) is unacceptable, the test is established as follows. Producer's risk is the probability of rejection although 30 min is the median, and consumer's risk is the probability of acceptance although 30 min is only the 25th percentile. Let these both equal 10%. Equations then give the value of sample size as 23 and the criterion as 14. Hence if more than 14 of the observed values exceed 30 min the test is failed.

Test Method 4

The method tests the median time. The median is the value, in any distribution, such that 50% of values exceed it and 50% do not. Only in the normal distribution does the median equal the mean. A log normal distribution is assumed in this test which has a fixed sample size of 20. The test involves comparing log MTTR in the test with log of the median value required in a given equation.

Test Method 5

The method tests the 'Chargeable Down Time per Flight'. This means the down time attributable to failures as opposed to passive maintenance activities, test-induced failures, modifications, etc. It is distribution free with a minimum sample size of 50 and can be used, indirectly, to demonstrate availability.

Test Method 6

The method is applicable to aeronautical systems and tests the 'Man-hour Rate'. This is defined as

$$\frac{\text{Total Chargeable Maintenance Man-hours}}{\text{Total Demonstration Flight Hours}}$$

Actual data are used and no consumer or producer risks apply.

Test Method 7

This is similar to Test Method 6 and tests the man-hour rate for simulated faults. There is a minimum sample size of 30.

Test Methods 1 – 4 are of a general nature whereas methods 5 – 7 have been developed with aeronautical systems in mind. In applying any test the risks must be carefully evaluated. There is a danger, however, of attaching an importance to results in proportion to the degree of care given to the calculations. It should therefore be emphasized that attention to the items listed in Section 15.2.1 in order to ensure that they reflect the agreed maintenance environment is of equal if not greater importance.

15.2.3 Data collection

It would be wasteful to regard the demonstration test as no more than a means of determining compliance with a specification. Each repair is a source of maintainability design evaluation and a potential input to the manual. Diagnostic instructions should not be regarded as static but be updated as failure information accrues. If the feedback is to be of use it is necessary to record each repair with the same detail as is called for in field reporting. The different repair elements of diagnosis, replacement, access, etc. should be separately listed, together with details of tools and equipment used. Demonstration repairs are easier to control than field maintenance and should therefore be better documented.

In any maintainability (or reliability) test the details should be fully described in order to minimize the possibilities of disagreement. Both parties should understand fully the quantitative and qualitative risks involved.

16 Quantified reliability centred maintenance

16.1 WHAT IS QRCM?

Quantitative Reliability Centred Maintenance (QRCM) involves calculations to balance the cost of excessive maintenance against that of the unavailability arising from insufficient maintenance. The following example illustrates one of the techniques which will be dealt with in this chapter.

Doubling the proof-test interval of a shutdown system on an off-shore production platform might lead to an annual saving of 2 man-days (say £2000). The cost in increased production unavailability might typically be calculated as 8×10^{-7} in which case the loss would be $8 \times 10^{-7} \times$ say £10 (per barrel) $\times$ say 50K (barrels) $\times$ 365 (days) = £146. In this case the reduction in maintenance is justified as far as cost is concerned.

QRCM is therefore the use of reliability techniques to optimize:

- Replacement (discard) intervals
- Spares holdings
- Proof-test intervals
- Condition monitoring

The first step in planning any QRCM strategy is to identify the critical items affecting plant unavailability since the greater an item's contribution to unavailability (or hazard) the more potential savings are to be made from reducing its failure rate.

Reliability modelling techniques lend themselves to this task in that they allow comparative availabilities to be calculated for a number of maintenance regimes. In this way the costs associated with changes in maintenance intervals, spares holdings and preventive replacement (discard) times can be compared with the savings achieved.

An important second step is to obtain site specific failure data. Although QRCM techniques can be applied using GENERIC failure rates and down times, there is better precision from site specific data. This is not, however, always available and published data sources (such as FARADIP.THREE) may have to be used. These are described in Chapter 4.

Because of the wide range of generic failure rates, plant specific data is preferred and an accurate plant register goes hand in hand with this requirement. Plant registers are often out of date and should be revised at the beginning of a new QRCM initiative. Thought should be given to a rational, hierarchical numbering for plant items which will assist in sorting like items, related items and items with like replacement times for purposes of maintenance and spares scheduling.

Good data is essential because, in applying QRCM, it is vital to take account of the way in which failures are distributed with time. We need to know if the failure rate is constant or whether it is increasing or decreasing. Preventive replacement (discard), for example, is only justified if there is an increasing failure rate.

16.2 THE QRCM DECISION PROCESS

The use of these techniques depends upon the failure distribution, the degree of redundancy and whether the cost of the maintenance action is justified by the saving in operating costs, safety or environmental impact. Figure 16.1 is the QRCM decision algorithm which can be used during

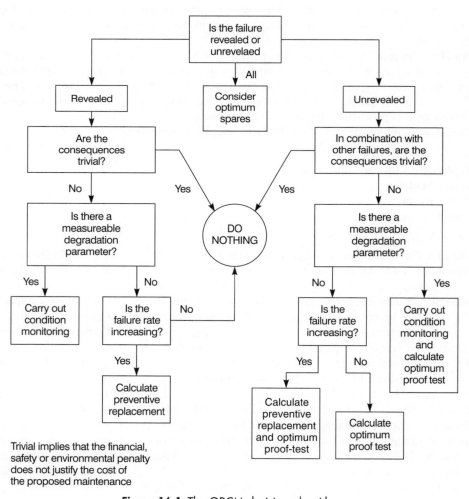

Trivial implies that the financial,
safety or environmental penalty
does not justify the cost of
the proposed maintenance

Figure 16.1 The QRCM decision algorithm

FMEA. As each component failure is considered the QRCM algorithm provides the logic which leads to the use of each of the techniques.

Using Figure 16.1 consider an unrevealed failure which, if it coincides with some other failure, leads to significant consequences such as the shutdown of a chemical plant. Assume that there is no measurable check whereby the failure can be pre-empted. Condition monitoring is not therefore appropriate. Assume, also, that the failure rate is not increasing therefore preventive discard cannot be considered. There is, however, an optimum proof-test interval whereby the cost of proof-test can be balanced against the penalty cost of the coincident failures.

16.3 OPTIMUM REPLACEMENT (DISCARD)

Specific failure data is essential for this technique to be applied sensibly. There is no generic failure data describing wearout parameters which would be adequate for making discard decisions. Times to failure must be obtained for the plant items in question and the Weibull techniques described in Chapter 6 applied. Note that units of time may be hours, cycles, operations or any other suitable base.

Only a significant departure of the shape parameter from ($\beta = 1$) justifies considering discard.

If $\beta \leq 1$ then there is no justification for replacement or even routine maintenance. If, on the other hand, $\beta > 1$ then there may be some justification for considering a preventive replacement before the item has actually failed. This will only be justified if the costs associated with an unplanned replacement (due to failure) are greater than those of a planned discard/replacement.

If this is the case then it is necessary to calculate:

(a) the likelihood of a failure (i.e. $1 - \exp(-t/\eta)^\beta$) in a particular interval times the cost of the unplanned failure.
(b) the cost of planned replacements during that interval.

The optimum replacement interval which minimizes the sum of the above two costs can then be found. Two maintenance philosophies are possible:

- Age replacement
- Block replacement

For the Age replacement case, an interval starts at time $t = 0$ and ends either with a failure or with a replacement at time $t = T$, whichever occurs first. The probability of surviving until time $t = T$ is $R(T)$ thus the probability of failing is $1 - R(T)$. The average duration of all intervals is given by:

$$\int_0^T R(t) \, dt$$

Thus the cost per unit time is:

$$\frac{[£_u \times (1 - R(T)) + £_p \times R(T)]}{\displaystyle\int_0^T R(t)\, dt}$$

where $£_u$ is the cost of unplanned outage (i.e. failure) and $£_p$ is the cost of a planned replacement.

For the Block replacement case, replacement always occurs at time $t = T$ despite the possibility of failures occurring before time $t = T$. For this case the cost per unit time is:

$$(£_u \times T)/\text{MTBF} \times T + £_p/T = £_u/\text{MTBF} + £_p/T$$

Note that, since the failure rate is not constant ($\beta > 1$), the MTBF used in the formula varies as a function of T.

There are two maintenance strategies involving preventive replacement (discard):

(a) If a failure occurs replace it and then wait the full interval before replacing again. This is known as AGE replacement.
(b) If a failure occurs replace it and nevertheless replace it again at the expiration of the existing interval. This is known as BLOCK replacement.

AGE replacement would clearly be more suitable for expensive items whereas BLOCK replacement might be appropriate for inexpensive items of which there are many to replace. Furthermore, BLOCK replacement is easier to administer since routine replacements then occur at regular intervals.

The COMPARE software package calculates the replacement interval for both cases and such that the sum of the following two costs is minimized:

● The cost of Unplanned replacement taking account of the likelihood that it will occur.
   PLUS
● The cost of the Scheduled replacement.

The program requests the Unplanned and Planned maintenance costs as well as the SHAPE and SCALE parameters.

Clearly the calculation is not relevant unless:

● SHAPE parameter, $\beta > 1$
   AND
● Unplanned Cost > Planned Cost

COMPARE provides a table of total costs (for the two strategies) against potential replacement times as can be seen in the following table where 1600 hours (nearly 10 weeks) is the optimum. It can be seen that the Age and Block replacement cases do not yield quite the same cost per unit time and that Block replacement is slightly less efficient. The difference may, however, be more than compensated for by the savings in the convenience of replacing similar items at the same time. Chapter 6 has already dealt with the issue of significance and of mixed failure modes.

Shape parameter (Beta) = 2.500
Scale parameter (Eta)   = 4000 hours

Cost of unscheduled replacement = £4000
Cost of planned replacement   = £500

Replacement interval	Cost per unit time	
	Age replace	Block replace
1000.	0.6131	0.6234
1200.	0.5648	0.5777
1400.	0.5429	0.5582
1600.	0.5381	0.5554
1800.	0.5451	0.5637
2000.	0.5605	0.5796
2200.	0.5820	0.6006
2400.	0.6080	0.6250
2600.	0.6372	0.6515
2800.	0.6688	0.6789
3000.	0.7018	0.7064

16.4 OPTIMUM SPARES

There is a cost associated with carrying spares, namely capital depreciation, space, maintenance, etc. In order to assess an optimum spares level it is necessary therefore to calculate the unavailability which will occur at each level of spares holding. This will depend on the following variables:

- Number of spares held
- Failure rate of the item
- Number of identical items in service
- Degree of redundancy within those items
- Lead time of procurement of spares
- Replacement (Unit Down Time) time when an item fails

This relationship can be modelled by means of Markov state diagram analysis and was fully described in Chapter 14 (Section 14.2.6).

It should be noted that, as the number of spares increases, there is a diminishing return in terms of improved unavailability until the so called 'infinite spares' case is reached. This is where the unavailability is dominated by the repair time and thus increased spares holding becomes ineffectual. At this point, only an improvement in repair time or in failure rate can increase the availability.

The cost of unavailability can be calculated for, say, zero spares. The cost saving in reduced unavailability can then be compared with the cost of carrying one spare and the process repeated until the optimum spares level is assessed.

The COMPARE package allows successive runs to be made for different spares levels. Figure 14.5 shows the Markov state diagram for 4 units with up to 2 spares.

16.5 OPTIMUM PROOF-TEST

In the case of redundant systems where failed redundant units are not revealed then the option of periodic proof-test arises. Although the failure rate of each item is constant, the system failure rate actually increases.

The Unavailability of a system can be calculated using the methods described in Chapter 8. It is clearly dependent partly on the proof-test interval which determines the down time of a failed (dormant) redundant item.

The technique involves calculating an optimum proof-test interval for revealing dormant failures. It seeks to trade off the cost of the proof-test (i.e. preventive maintenance) against the reduction in unavailability.

It applies where coincident dormant failures cause unavailability. An example would be the failure to respond of both a 'high' alarm and a 'high high' signal.

The unavailability is a function of the instrument failure rates and the time for which dormant failures persist. The more frequent the proof test, which seeks to identify the dormant failures, then the shorter is the down time of the failed items.

Assume that the 'high' alarm and 'high high' signal represent a duplicated redundant arrangement. Thus, one instrument may fail without causing plant failure (shutdown).

It has already been shown that the reliability of the system is given by:

$$R(t) = 2\,e^{-\lambda t} - e^{-2\lambda t}$$

Thus the probability of failure is $1 - R(t)$

$$= 1 - 2\,e^{-\lambda t} + e^{-2\lambda t}$$

If the cost of an outage (i.e. lost production) is $£_u$ then the expected cost, due to outage, is:

$$= (1 - 2\,e^{-\lambda t} + e^{-2\lambda t}) \times £_u$$

Now consider the proof test, which costs $£_p$ per visit. If the proof test interval is T then the expected cost, due to preventive maintenance, is:

$$= (2\,e^{-\lambda t} - e^{-2\lambda t}) \times £_p$$

The total cost per time interval is thus:

$$= [(1 - 2\,e^{-\lambda t} + e^{-2\lambda t}) \times £_u] + [(2\,e^{-\lambda t} - e^{-2\lambda t}) \times £_p]$$

The average length of each interval is $\int_0^T R(t)\,dt$

$$= 3/2\lambda - 2/\lambda\,e^{-\lambda T} + 1/2\lambda\,e^{-2\lambda T}$$

The total cost per unit time can therefore be obtained by dividing the above expression into the preceding one.

The minimum cost can be found by tabulating the cost against the proof-test interval (*T*). In the general case the total cost per unit time is:

$$= \frac{[(1 - R(T)) \times \pounds_{\mathrm{u}}] + [R(T) \times \pounds_{\mathrm{p}}]}{\displaystyle\int_0^T R(t)\mathrm{d}t}$$

Again, the COMPARE package performs this calculation and provides an optimum interval (approximately 3 years) as can be seen in the following example.

Total number of units	= 2
Number of units required	= 1
MTBF of a single unit	= 10.00 years
Cost of unscheduled outage	= £2000
Cost of a planned visit	= £100

Proof-test interval	Cost per unit time
1.000	117.6
1.700	86.88
2.400	78.98
3.100	77.79
3.800	79.18
4.500	81.65
5.200	84.56
5.900	87.60
6.600	90.61
7.300	93.51
8.000	96.28

16.6 CONDITION MONITORING

Many failures do not actually occur spontaneously but develop over a period of time. It follows, therefore, that if this gradual 'degradation' can be identified it may well be possible to pre-empt the failure. Overhaul or replacement are then realistic options. During the failure mode analysis it may be possible to determine parameters which, although not themselves causing a hazard or equipment outage, are indicators of the degradation process.

In other words, the degradation parameter can be monitored and action taken to prevent failure by observing trends. Trend analysis would be carried out on each of the measurements in order to determine the optimum point for remedial action.

It is necessary for there to be a reasonable time period between the onset of the measurable degradation condition and the actual failure. The length (and consistency) of this period will determine the optimum inspection interval.

There are a number of approaches to determining the inspection interval. Methods involving a gradual increase in interval run the risk of suffering the failure. This may be expensive or hazardous. Establishing the interval by testing, although safer, is expensive, may take time and relies on simulated operating environments. However, in practice, a sensible mixture of experience and data can lead to realistic intervals being chosen. By concentrating on a specific failure mode (say valve diaphragm leakage) and by seeking out those with real operating experience it is possible to establish realistic times. Even limited field and test data will enhance the decision.

The following list provides some examples of effects which can be monitored:

- regular gas and liquid emission leak checks
- critical instrumentation parameter measurements (gas, fire, temp, level, etc.)
- insulation resistivity
- vibration measurement and analysis of characteristics
- proximity analysis
- shock pulse monitoring
- acoustic emission
- corrosive states (electro-chemical monitoring)
- dye penetration
- spectrometric oil analysis
- electrical insulation
- hot spots
- surface deterioration
- state of lubrication and lubricant
- plastic deformation
- balance and alignment.

17 Software quality/reliability

17.1 PROGRAMMABLE DEVICES

There has been a spectacular growth since the 1970s in the use of programmable devices. They have made a significant impact on methods of electronic circuit design. The main effect has been to reduce the number of different circuit types by the use of computer architecture. Coupled with software programming, this provides the individual circuit features previously achieved by differences in hardware. The word 'software' refers to the instructions needed to enable a programmable device to function, including the associated hierarchy of documents required to produce that code. This use of programming at the circuit level, now common with most industrial and consumer products, brings with it some associated quality and reliability problems. When applied to microprocessors at the circuit level the programming, which is semi-permanent and usually contained in ROM (Read Only Memory), is known as Firmware. The necessary increase in function density of devices in order to provide the large quantities of memory in small packages has matched this trend.

Computing and its associated software is seen in three broad categories:

1. *Mainframe computing*: This can best be visualized in terms of systems which provide a very large number of terminals and support a variety of concurrent tasks. Typical functions are interactive desktop terminals or bank terminals. Such systems are also characterized by the available disk and tape storage which often runs into hundreds of megabytes.
2. *Minicomputing*: Here we are dealing with a system whose CPU may well deal with the same word length (32 bits) as the mainframe. The principal difference lies in the architecture of the main components and, also, in the way in which it communicates with the peripherals. A minicomputer can often be viewed as a system with a well-defined hardware interface to the outside world enabling it to be used for process monitoring and control.
3. *Microprocessing*: The advent of the microcomputer is relatively recent but it is now possible to have a 32-bit architecture machine as a desktop computer. These systems are beginning to encroach on the minicomputer area but are typically being used as 'personal computers' or as sophisticated workstations for programming, calculating, providing access to mainframe, and so on.

The boundaries between the above categories have blurred considerably in recent years to the extent that minicomputers now provide the mainframe performance of a few years ago. Similarly, microcomputers provide the facilities expected from minis.

From the quality and reliability point of view, there are both advantages and disadvantages arising from programmable design solutions:

Reliability advantages	Reliability disadvantages
Less hardware (fewer devices) per circuit.	Difficult to 'inspect' software for errors.
Fewer device types.	Difficult to impose standard approaches
Consistent architecture (configuration).	to software design.
Common approach to hardware design.	Difficult to control software changes.
Easier to support several models (versions) in the field.	Testing of LSI devices difficult owing to high package density and therefore
Simpler to modify or reconfigure.	reduced interface with test equipment.
	Impossible to predict software failures.

17.2 SOFTWARE FAILURES

The question arises as to how a software failure is defined. Unlike hardware, there is no physical change associated with a unit that is 'functional' at one moment and 'failed' at the next. Software failures are in fact errors which, owing to the complexity of a computer program, do not become evident until the combination of conditions brings the error to light. The effect is then the same as any other failure. Unlike the hardware Bathtub, there is no wearout characteristic but only a continuing burn-in. Each time that a change to the software is made the error rate is likely to rise, as shown in Figure 17.1. As a result of software errors there has been, for some time, an interest in developing methods of controlling the activities of programmers and of reducing software complexity by attempts at standardization.

Figure 17.2 illustrates the additional aspect of software failures in programmable systems. It introduces the concept of Fault/Error/Failure. Faults may occur in both hardware and software. Software faults, often known as bugs, will appear as a result of particular portions of code being used for the first time under a particular set of circumstances.

The presence of a fault in a programmed system does not necessarily result in either an error or a failure. A long time may elapse before that code is used under the circumstances which lead to failure.

A fault (bug) may lead to an error, which occurs when the system reaches an incorrect state. That is, a bit, or bits, takes an incorrect value in some store or highway.

An error may propagate to become a failure if the system does not contain error-recovery software capable of detecting and eliminating the error.

Failure, be it for hardware or software reasons, is the termination of the ability of an item to perform the function specified.

Figure 17.1 Software error curve

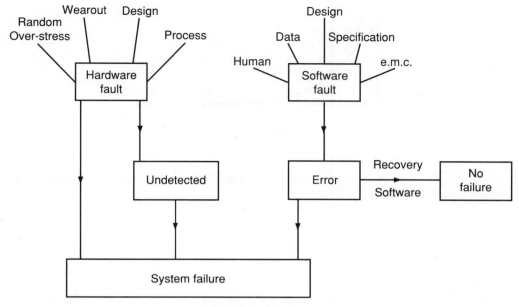

Figure 17.2

It should be understood that the term 'software' refers to the complete hierarchy of documentation which defines a programmable system. This embraces the Requirements Specification, Data Specifications, Subsystem Specifications and Module definitions, as well as the Flowcharts, Listings and Media which are often thought of as comprising the entire software.

Experience shows that less than 1% of software failures result from the actual 'production' of the firmware. This is hardly surprising since the act of inputting code is often self-checking and errors are fairly easy to detect. This leaves the design and coding activities as the source of failures. Within these, less than 50% of errors are attributed to the coding activity. Software reliability is therefore inherent in the design process of breaking down the requirements into successive levels of specification.

17.3 SOFTWARE FAILURE MODELLING

Numerous attempts have been made to design models which enable software failure rates to be predicted from the initial failures observed during integration and test or from parameters such as the length and nature of the code. The latter suffers from the difficulty that, in software, there are no elements (as with hardware components) with failure characteristics which can be taken from experience and used for predictive purposes. This type of prediction is therefore unlikely to prove successful. The former method, of modelling from the early failures, suffers from a

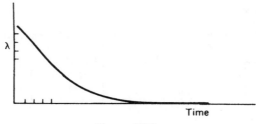

Figure 17.3

difficulty which is illustrated by this simple example. Consider the following failure pattern based on 4 days of testing:

Day 1	10 failures
Day 2	9 failures
Day 3	8 failures
Day 4	7 failures

To predict, from these data, when 6 failures per day will be observed is not difficult, but what is required is to know when the failure rate will be 10^{-4} or perhaps 10^{-5}. It is not certain that the information required is in fact contained within the data. Figure 17.3 illustrates the coarseness of the data and the fact that the tail of the distribution is not well defined and by no means determined by the shape of the left-hand end.

A number of models have been developed. They rely on various assumptions concerning the nature of the failure process, such as the idea that failure rate is determined by the number of potential failures remaining in the program. These are by no means revealed solely by the passage of calendar time, since repeated executions of the same code will not usually reveal further failures.

Present opinion is that no one model is better than any other, and it must be said that, in any case, an accurate prediction only provides a tool for scheduling rather than a long-term field reliability assessment. The models include:

- *Jelinski Moranda*: This assumes that failure rate is proportional to the remaining fault content. Remaining faults are assumed to be equally likely to occur.
- *Musa*: Program execution rather than calendar time is taken as the variable.
- *Littlewood Verall*: Assumes successive execution time between failures to be an exponentially distributed random variable.
- *Structured Models*: These attempt to break software into subunits. Rules for switching between units and for the failure rate of each unit are developed.
- *Seeding and Tagging*: This relies on the injection of known faults into the software. The success rate of debugging of the known faults is used to predict the total population of failures by applying the ratio of success to the revealed non-seeded failures. For this method to be successful one has to assume that the seeded failures are of the same type as the unknown failures.

Clearly, the number of variables involved is large and their relationship to failure rate far from precise. It is the author's view that, currently, qualitative activities in Software Quality Assurance are more effective than attempts at prediction.

17.4 SOFTWARE QUALITY ASSURANCE

Software QA, like hardware QA, is aimed at preventing failures. It is based on the observation that software failures are predominantly determined by the design. Experience in testing real-time software controlled systems shows that 50% of software 'bugs' result from unforeseen combinations of real-time operating events which the program instructions cannot accommodate. As a result, the algorithm fails to generate a correct output or instruction and the system fails.

Software QA is concerned with:

Organization of Software QA effort (Section 17.4.1)
Documentation Controls (17.4.2)
Programming Standards (17.4.3)
Design Features (17.4.4)
Code Inspections and Walkthroughs (17.4.5)
Integration and Test (17.4.6)

The following sections outline these areas and this chapter concludes with a number of checklist questions suitable for audit or as design guidelines.

17.4.1 Organization of Software QA

There needs to be an identifiable organizational responsibility for Software QA. The important point is that the function can be identified. In a small organization, individuals often carry out a number of tasks. It should be possible to identify written statements of responsibility for Software QA, the maintenance of standards and the control of changes.

There should be a Quality Manual, Quality Plans and specific Test Documents controlled by QA independently of the project management. They need not be called by those names and may be contained in other documents. It is the intent which is important. Main activities should include:

Configuration Control
Library of Media and Documentation
Design Review
Auditing
Test Planning

17.4.2 Documentation controls

There must be an integrated hierarchy of Specification/Documents which translates the functional requirements of the product through successive levels of detail to the actual source code. In the simplest case this could be satisfied by:

A Functional description and
A Flowchart or set of High-Level Statements and
A Program listing.

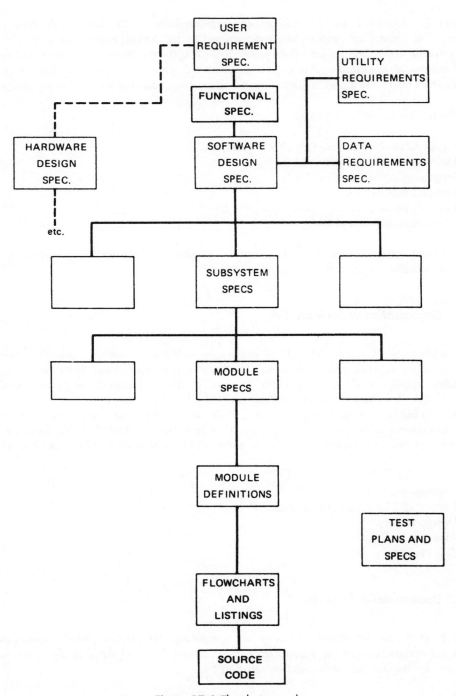

Figure 17.4 The design cycle

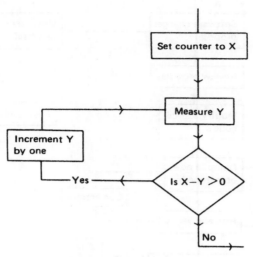

Figure 17.5 Flowchart

In more complex systems there should be a documentation hierarchy. The design must focus onto a user requirements specification which is the starting point in a top-down approach.

In auditing software it is important to look for such a hierarchy and to establish a diagram similar to Figure 17.4, which reflects the product, its specifications and their numbering system. Failure to obtain this information is a sure indicator that software is being produced with less than adequate controls. Important documents are:

- *User requirements specification*: Describes the functions required of the system. It should be unambiguous and complete and should describe what is required and not how it is to be achieved. It should be quantitative, where possible, to facilitate test planning. It states *what* is required and must not pre-empt and hence constrain the design.
- *Functional specification*: Whereas the User requirements specification states *what* is required, the Functional specification outlines *how* it will be achieved. It is usually prepared by the developer in response to the requirements.
- *Software design specification*: Takes the above requirements and, with regard to the hardware configuration, describes the functions of processing which are required and addresses such items as language, memory requirements, partitioning of the program into accessible subsystems, inputs, outputs, memory organization, data flow, etc.
- *Subsystem specification*: This should commence with a brief description of the subsystem function. Interfaces to other subsystems may be described by means of flow diagrams.
- *Module specification*: Treating the module as a black box, it describes the interfaces with the rest of the system and the functional performance as perceived by the rest of the software.
- *Module definition*: Describes the working of the software in each module. It should include the module test specification, stipulating details of input values and the combinations which are to be tested.
- *Charts and diagrams*: A number of techniques are used for charting or describing a module. The most commonly known is the flowchart, shown in Figure 17.5. There are, however, alternatives, particularly in the use of high-level languages. These involve diagrams and pseudo-code.

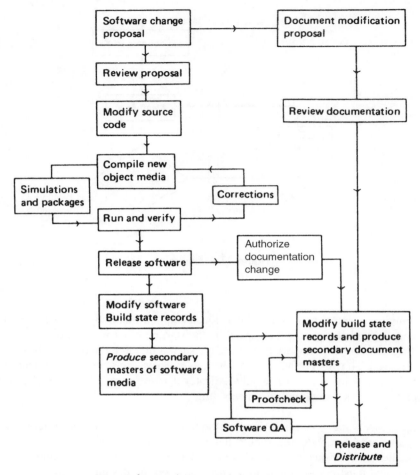

Figure 17.6 Software change and documentation procedure

- *Utilities specification*: This should contain a description of the hardware requirements, including the operator interface and operating system, the memory requirements, processor hardware, data communications and software support packages.
- *Development notebooks*: An excellent feature is the use of a formal development notebook. Each designer opens a loose-leaf file in which is kept all specifications, listings, notes, change documentation and correspondence pertaining to that project.

Change control

As with hardware, the need to ensure that changes are documented and correctly applied to all media and program documents is vital. All programs and their associated documents should therefore carry issue numbers. A formal document and software change procedure is required (see Figure 17.6) so that all change proposals are reviewed for their effect on the total system.

17.4.3 Programming standards

The aim of structured programming is to reduce program complexity by using a library of defined structures wherever possible. The human brain is not well adapted to retaining random information and sets of standard rules and concepts substantially reduce the likelihood of error. A standard approach to creating files, polling output devices, handling interrupt routines, etc. constrains the programmer to use the proven methods. The use of specific subroutines is a further step in this direction. Once a particular sequence of program steps has been developed in order to execute a specific calculation, then it should be used as a library subroutine by the rest of the team. Re-inventing the wheel is both a waste of time and an unnecessary source of failure if an error-free program has already been developed.

A good guide is 30–60 lines of coding plus 20 lines of comment. Since the real criterion is that the module shall be no larger than to permit a total grasp of its function (that is, it is perceivable), it is likely that the optimum size is a line print page (3 at most).

The use of standard sources of information is of immense value. Examples are:

> Standard values for constants
> Code Templates (standard pieces of code for given flowchart elements)
> Compilers

The objective is to write clear, structured software, employing well-defined modules whose functions are readily understood. There is no prize for complexity.

There are several methods of developing the module on paper. They include:

> Flow Diagrams
> Hierarchical Diagrams
> Structured Box Diagrams
> Pseudo-code

17.4.4 Fault-tolerant design features

Fault Tolerance can be enhanced by attention to a number of design areas. These features include:

- Use of redundancy, which is expensive. The two options are Dual Processing and Alternate Path (Recovery Blocks).
- Use of error-checking software involving parity bits or checksums together with routines for correcting the processing.
- Timely display of fault and error codes.
- Generous tolerancing of timing requirements.
- Ability to operate in degraded modes.
- Error confinement. Programming to avoid error proliferation or, failing that, some form of recovery.
- Watchdog timer techniques involve taking a feedback from the microprocessor and, using that clocked rate, examining outputs to verify that they are dynamic and not stuck in one state. The timer itself should be periodically reset.

- Faults in one microprocessor should not be capable of affecting another. Protection by means of buffers at inputs and outputs is desirable so that a faulty part cannot pull another part into an incorrect state. Software routines for regular checking of the state (high or low) of each part may also be used.
- Where parts of a system are replicated the use of separate power supplies can be considered, especially since the power supply is likely to be less reliable than the replicated processor.

17.4.5 Reviews

There are two approaches to review of code:

1. Code Inspection where the designer describes the overall situation and the module functions to the inspection team. The team study the documentation and, with the aid of previous fault histories, attempt to code the module. Errors are sought and the designer then carries out any rework, which is then re-inspected by the team.
2. The Structured Walkthrough in which the designer explains and justifies each element of code until the inspection team is satisfied that they agree and understand each module.

17.4.6 Integration and test

There are various types of testing which can be applied to software:

- *Dynamic Testing*: This involves executing the code with real data and I/O. At the lowest level this can be performed on development systems as is usually the case with Module Testing. As integration and test proceeds, the dynamic tests involve more of the actual equipment until the functional tests on the total equipment are reached. Aids to dynamic testing include automatic test beds and simulators which are now readily available. dynamic testing includes:
- *Path Testing*: This involves testing each path of the software. In the case of flowcharted design there are techniques for 'walking through' each path and determining a test. It is difficult, in a complex program, to be sure that all combinations have been checked. In fact the number of combinations may be too high to permit all paths to be tested.
- *Software Proving by Emulation*: An 'intelligent' communications analyser or other simulator having programmable stimulus and response facilities is used to emulate parts of the system not yet developed. In this way the software can be made to interact with the emulator which appears as if it were the surrounding hardware and software. Software testing can thus proceed before the total system is complete.
- *Functional Testing*: The ultimate empirical test is to assemble the system and to test every possible function. This is described by a complex test procedure and should attempt to cover the full range of environmental conditions specified.
- *Load Testing*: The situation may exist where a computer controls a number of smaller microprocessors, data channels or even hard-wired equipment. The full quantity of these peripheral devices may not be available during test, particularly if the system is designed for expansion. In these cases, it is necessary to simulate the full number of inputs by means of a simulator. A further micro or minicomputer may well be used for this purpose. Test software will then have to be written which emulates the total number of devices and sends and receives data from the processor under test.

Be most suspicious of repeated slip in a test programme. This is usually a symptom that the test procedure is only a cover for debug. Ideally, a complete error-free run of the test procedure is needed after debug, although this is seldom achieved in practice with large systems.

The practice of pouring in additional personnel to meet the project schedule is ineffective. The division of labour, below module level, actually slows down the project.

17.5 MODERN/FORMAL METHODS

The traditional Software QA methods, described in the previous section, are essentially open-ended checklist techniques. They have been developed over the last 15 years but would be greatly enhanced by the application of more formal and automated methods. The main problem with the existing open-ended techniques is that they provide no formal measures as to how many of the hidden errors have been revealed.

The term Formal Methods is much used and much abused. It covers a number of methodologies and techniques for specifying and designing systems, both non-programmable and programmable. They can be applied throughout the life-cycle including the specification stage and the software coding itself.

The term is used here to describe a range of mathematical notations and techniques applied to the rigorous definition of system requirements which can then be propagated into the subsequent design stages. The strength of formal methods is that they address the requirements at the beginning of the design cycle. One of the main benefits of this is that formalism applied at this early stage may lead to the prevention, or at least early detection, of incipient errors. The cost of errors revealed at this stage is dramatically less than if they are allowed to persist until commissioning or even field use. This is because the longer they remain undetected the more serious and far reaching are the changes required to correct them.

Figure 17.7 The quality problem

The three major quality problems, with software, are illustrated in Figure 17.7. First, the statement of requirements is in free language and thus the opportunity for ambiguity, error and omission is at a maximum. The very free language nature of the requirements makes it impossible to apply any formal or mathematical review process at this stage. It is well known that the majority of serious software failures originate in this part of the design cycle. Second, the source code, once produced, can only be reviewed by open-ended techniques as described in Section 17.4.4. Again, the discovery of ten faults gives no clue as whether one, ten or 100 remain. Third, the use of the software (implying actual execution of the code) is effectively a very small sample of the execution paths and input/output combinations which are possible in a typical piece of real-time software. Functional test is, thus, only a small contribution to the validation of a software system.

In these three areas of the design cycle there are specific developments:

- *Requirements Specification and Design*: There is emerging a group of design languages involving formal graphical and algebraic methods of expression. For requirements, such tools as VDM (Vienna Development Method), OBJ (Object Oriented Code) and Z (a method developed at Oxford University) are now in use. They require formal language statements and, to some extent, the use of Boolean expressions. The advantage of these methods is that they substantially reduce the opportunity for ambiguity and omission and provide a more formal framework against which to validate the requirements.

  Especial interest in these methods has been generated in the area of safety-related systems in view of their potential contribution to the safety integrity of systems in whose design they are used.

  The potential benefits are considerable but they cannot be realized without properly trained people and appropriate tools. Formal methods are not easy to use. As with all languages, it is easier to read a piece of specification than it is to write it. A further complication is the choice of method for a particular application. Unfortunately, there is not a universally suitable method for all situations.

  Formal methods are equally applicable to the design of hardware and software. In fact they have been successfully used in the design of large scale integration electronic devices as, for example the Viper chip produced by RSRE in Malvern, UK.

  It should always be borne in mind that establishing the correctness of software, or even hardware, alone is no guarantee of correct system performance. Hardware and software interact to produce a system effect and it is the specification, design and validation of the system which matters. This system-wide view should also include the effects of human beings and the environment.

  The potential for creating faults in the specification stage arises largely from the fact that it is carried out mainly in natural language. On one hand this permits freedom of expression and comprehensive description but, on the other, leads to ambiguity, lack of clarity and little protection against omission. The user communicates freely in this language which is not readily compatible with the formalism being suggested here.

- *Static Analysis*: This involves the algebraic examination of source code (not its execution). Packages are available (such as MALPAS from Fluor Global Services at Farnham, Surrey) which examine the code statements for such features as:

  The graph structure of the paths
  Unreachable code
  Use of variables
  Dependency of variables upon each other
  Actual semantic relationship of variables

Consider the following piece of code:
BEGIN
INTEGER A, B, C, D, E
A: = 0
NEXT: INPUT C:
IF C<0 THEN GOTO EXIT:
B: = B+C
D: = B/A
GOTO NEXT:
PRINT B, D;
EXIT: END;
Static analysis will detect that:

i)   B is not initialized before use.

ii) E is never used

iii) A is zero and is used as a divisor

iv) The PRINT B, D; command is never used because of the preceding statement.

Static analysis is extremely powerful in that it enables the outputs of the various analysers to be compared with the specification in order to provide a formal review loop between code and specification. A further advantage is that static analysis forces the production of proper specifications since they become essential in order to make use of the analyser outputs.

Figure 17.8 shows the packages of MALPAS (one such static analysis tool). It acts on the source code and Control flow analysis identifies the possible entry and exit points to the module, pieces of unreachable code and any infinitely looping dynamic halts. It gives an initial feel for the structure and quality of the program. Data use analysis identifies all the inputs and outputs of the module and checks that data is being correctly handled. For example, it checks that each variable is initialized before being used. Information flow analysis deduces the information on which each output depends. The Path assessor is used to provide a measure of the complexity in that the number of paths through the code is reported for each procedure. Semantic analysis identifies the actions taken on each feasible path through a procedure. In

Figure 17.8 The MALPAS suite

particular, it rewrites imperative, step-by-step procedures into a declarative, parallel assignment form. The analyst can use this to provide an alternative perspective on the function of the procedure. The result of the analyser is to tell the analyst the actual relationship of the variables to each other. Compliance analysis attempts to prove that a procedure satisfies a specified condition. For example, it could be used to check that the result of the procedure 'sort' is a sequence of items where each item is bigger than the preceding one. The report from the compliance analysis identifies those input values for which the procedure will fail.

- *Test Beds*: During dynamic testing (involving actual execution of the code), automated 'test beds' and 'animators' enable testing to proceed with the values of variables being displayed alongside the portions of code under test. Numerous test 'tools' and so-called 'environments' are commercially available and continue to be developed.

17.6 SOFTWARE CHECKLISTS

17.6.1 Organization of Software QA

1. Is there a senior person with responsibility for Software QA and does he or she have adequate competence and authority to resolve all software matters?
2. Is there evidence of regular reviews of Software Standards?
3. Is there a written company requirement for the Planning of a Software Development?
4. Is there evidence of Software Training?
5. Is there a Quality Manual or equivalent documents?
6. Is there a system for labelling all Software Media?
7. Is there a Quality Plan for each development including

> Organization of the team
> Milestones
> Codes of Practice
> QC procedures, including release
> Purchased Software
> Documentation Management
> Support Utilities
> Installation
> Test Strategy?

8. Is there evidence of documented design reviews? The timing is important. So-called reviews which are at the completion of test are hardly design reviews.
9. Is there evidence of defect reporting and corrective action?
10. Are the vendor's quality activities carried out by people not involved in the design of the product that they are auditing?
11. Is there a fireproof media and file store?
12. Are media duplicated and separately stored?

17.6.2 Documentation controls

1. Is there an adequate structure of documentation for the type of product being designed?
2. Do all the documents exist?
3. Do specifications define what must not happen as well as what must?

4. Is there a standard or guide for flowcharts, diagrams or pseudo-code in the design of modules?
5. Are there written conventions for file naming and module labelling?
6. Is there a person with specific responsibility for Documentation Control?
7. Is there a person with specific responsibility for Change Control?
8. Is there a distribution list for each document?
9. Are there established rules for the holding of originals?
10. Are all issues of program media accurately recorded?
11. Is there a system for the removal and destruction of obsolete documents from all work areas?
12. Are media containing non-conforming software segregated and erased?

17.6.3 Programming standards

1. Is there a library of common program modules?
2. Is the 'top-down' approach to Software Design in evidence?
3. Is high-level or low-level language used? Has there been a conscious justification?
4. Is there a document defining program standards?
5. Is there reference to Structured Programming?
6. Is each of the following covered?

> Block lengths
> Size of codable units (Module Size)
> Use of globals
> Use of GOTO statements
> File, Operator error, and Unauthorized use security
> Recovery conventions
> Data organization and structures
> Memory organization and backup
> Error-correction software
> Automatic fault diagnosis
> Range checking of arrays
> Use of PROM, EPROM, RAM, DISC, etc.
> Structured techniques
> Treatment of variables (that is, access)
> Coding formats
> Code layout
> Comments (REM statements)
> Rules for module identification.

17.6.4 Design features

1. Is there evidence that the following are taken into consideration?

> Electrical protection (mains, airborne)
> Power supplies and filters
> Opto isolation, buffers
> Earthing
> Battery backup
> Choice of processors

Use of language
Rating of I/O devices
Redundancy (dual programming)
Data communications
Human/machine interface
Layout of hardware
Hardware configuration (e.g. multidrops)
Watchdog timers
RAM checks
Error confinement
Error detection
Error recovery.

2. Are there syntax and protocol-checking algorithms?
3. Are interfaces defined such that illegal actions do not corrupt the system or lock up the interface?
4. Are all data files listed (there should be a separate list)?
5. Were estimates of size and timing carried out?
6. Are the timing criteria of the system defined where possible?
7. Will it reconstruct any records that may be lost?
8. Are there facilities for recording system state in the event of failure?
9. Have acceptable degraded facilities been defined?
10. Is there a capability to recover from random jumps resulting from interference?
11. Are the following adequate?

Electrical protection (mains and e.m.i.)
Power suppliers and filters
Earthing.

12. Is memory storage adequate for foreseeable expansion requirements?
13. Are data link lengths likely to cause timing problems?
14. Are the following suitable for the application in hand?

Processor
Peripherals
Operating System
Packaging.

15. Is there evidence of a hardware/software trade-off study?
17. Is use made of watchdog timers to monitor processors?

Coding formats
Code layout
Comments (REM statements)
Rules for module identification.

17.6.5 Code inspections and walkthroughs

1. Are all constants defined?
2. Are all unique values explicitly tested on input parameters?
3. Are values stored after they are calculated?

4. Are all defaults explicitly tested on input parameters?
5. If character strings are created are they complete? Are all delimiters shown?
6. If a parameter has many unique values, are they all checked?
7. Are registers restored on exits from interrupts?
8. Should any register's contents be retained when re-using that register?
9. Are all incremental counts properly initialized (0 or 1)?
10. Are absolute addresses avoided where there should be symbolics?
11. Are internal variable names unique or confusing if concatenated?
12. Are all blocks of code necessary or are they extraneous (e.g. test code)?
13. Are there combinations of input parameters which could cause a malfunction?
14. Can interrupts cause data corruption?
15. Is there adequate commentary (REM statements) in the listing?
16. Are there time or cycle limitations placed on infinite loops?

17.6.6 Integration and test

1. Are there written requirements for testing Subcontracted or Proprietary Software?
2. Is there evidence of test reporting and remedial action?
3. Is there evidence of thorough environmental testing?
4. Is there a defect-recording procedure in active use?
5. Is there an independent Test Manager appointed for the test phase of each development programme?
6. Is there a comprehensive system of test documentation (e.g. test plans, specifications, schedules) for each product?
7. Is there an effective system of calibration and control of test equipment?
8. Do test plans indicate a build-up of testing (e.g. module test followed by subsystem test followed by system test)?
9. Do test schedules permit adequate time for testing?
10. Is there evidence of repeated slip in the test programme?
11. To what extent are all the paths in the program checked?
12. Does the overall design of the tests attempt to prove that the system behaves correctly for improbable real-time events (e.g. Misuse tests)?

Note: This chapter is a brief summary of *Achieving Quality Software*,
    D. J. Smith, Chapman Hall, 1995 ISBN 0 412 6227 0 X.

Part Five
Legal, Management and Safety Considerations

18 Project management

18.1 SETTING OBJECTIVES AND SPECIFICATIONS

Realistic reliability and maintainability (RAM) objectives need to be set with due regard to the customer's design and operating requirements and cost constraints. In the case of contract development or plant engineering, these are likely to be outlined in a tender document or a requirements specification. Some discussion and joint study with the customer may be required to establish economic reliability values which sensibly meet his or her requirements and are achievable within the proposed technology at the costs allowed for. Over-specifying the requirement may delay the project when tests eventually show that objectives cannot be met and it is realized that budgets will be exceeded.

When specifying an MTBF it is a common mistake to state a confidence level; in fact the MTBF requirement stands alone. The addition of a confidence level implies a statistical demonstration and supposes that the MTBF would be established by a single demonstration at the stated confidence. On the contrary, a design objective is a target and must be stated without statistical limitations.

Vague statements such as 'high reliability' and 'the highest quality' should be avoided at all costs. They are totally subjective and cannot be measured. Therefore they cannot be demonstrated or proved.

Consideration of the equipment type and the use to which it is put will influence the parameters chosen. Remember the advice given in Chapter 2 about the meaning and applicability of failure rate, MTBF, Availability, MTTR, etc.

A major contribution to the problems associated with reliability and quality comes from the lack of (or inadequacy of) the engineering design specification. It should specify the engineering requirements in full, including reliability and MTTR parameters. These factors should include:

1. Functional description: speeds, functions, human interfaces and operating periods.
2. Environment: temperature, humidity, etc.
3. Design life: related to wearout and replacement policy.
4. Physical Parameters: size and weight restrictions, power supply limits.
5. Standards: BS, US MIL, Def Con, etc., standards for materials, components and tests.
6. Finishes: appearance and materials.
7. Ergonomics: human limitations and safety considerations.
8. Reliability, availability and maintainability: module reliability and MTTR objectives. Equipment R and M related to module levels.
9. Manufacturing quantity: Projected manufacturing levels – First off, Batch, Flow.
10. Maintenance philosophy: Type and frequency of preventive maintenance. Repair level, method of diagnosis, method of second-line repair.

18.2 PLANNING, FEASIBILITY AND ALLOCATION

The design and assurance activities described in this book simply will not take place unless there is real management understanding and committment to a reliability and maintainability programme with specific resources allocated. Responsibilities have to be placed on individuals for each of the activities and a reliability programme manager appointed with sufficient authority and the absence of conflicting priorities (that is, programme dates) to control the RAM objectives. Milestones, with dates, will be required against which progress can be measured as, for example:

> Completion of feasibility study (including RAM calculations).
> Reliability objectives for modules and for bought-out items allocated.
> Test specification prepared and agreed
> Prototype tests completed.
> Modifications arising from tests completed.
> Demonstrations of reliability and maintainability.
> Design review dates.

The purpose of a feasibility study is to establish if the performance specification can be met within the constraints of cost, technology, time and so on. This involves a brief reliability prediction, based perhaps on a block diagram approach, in order to decide if the design proposal has a reasonable chance of being engineered to meet the requirements. Allocation of objectives has been emphasized in Chapter 11 and is important if the objectives are not to be met by a mixture of over- and under-design.

It is useful to remember that there are three levels of RAM measurement:

PREDICTION: A modelling exercise which relies on the validity of historical failure rates to the design in question. This provides the lowest level of confidence.

STATISTICAL DEMONSTRATION TEST: This provides sample failure information (perhaps even zero failures in a given amount of time). It is usually in a test rather than field environment. Whilst providng more confidence than paper PREDICTION it is still subject to statistical risk and the limitations of a test environment.

FIELD DATA: Except in the case of very high reliability systems (e.g. submerged cable and repeater), realistic numbers of failures are obtained and can be used in a reliability growth programme as well as for comparison with the original targets.

18.3 PROGRAMME ACTIVITIES

The extent of the reliability and maintainability activities in a project will depend upon:

> The severity of the requirement.
> The complexity of the product.
> Time and cost constraints.
> Safety considerations.
> The number of items to be produced.

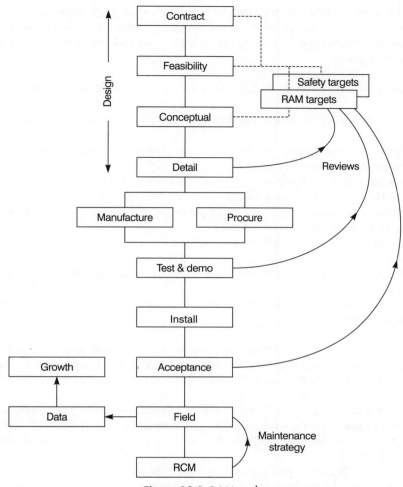

Figure 18.1 RAM cycle

A Safety and Reliability Plan must be produced for each project or development. Without this there is nothing against which to audit progress and, therefore, no formal measure of progress towards the targets. Figure 18.1 shows a simple RAM Design-Cycle which provides a model against which to view the activities. Figure 1.2, in Section 1.5, gave more detail.

These have all been covered in the book and include:

● Feasibility study – An initial 'prediction' to ascertain if the targets are realistic or impossible.
● Setting objectives – Discussed above with allocation and feasibility.
● Contract requirements – The formal agreement on the RAM targets, warranty, acceptance criteria, etc.

● Design Reviews – These are intended to provide an evaluation of the design at defined milestones. The design review team should include a variety of skills and be chaired by a person independent of the design team. The following checklist is a guide to the factors which might be considered:

1. Electrical factors involving critical features, component standards, circuit trade-offs, etc.
2. Software reliability including configuration control, flowcharts, user documentation, etc.
3. Mechanical features such as materials and finish, industrial design, ergonomics, equipment practice and so on.
4. Quality and reliability covering environmental testing, RAM predictions and demonstrations, FMECA, test equipment and procedures, trade-offs, etc.
5. Maintenance philosophy including repair policy, MTTR prediction, maintenance resource forecasts, customer training and manuals.
6. Purchased items involving lead times, multiple sourcing, supplier evaluation and make/buy decisions.
7. Manufacturing and installation covering tolerances, burn-in, packaging and transport, costs, etc.
8. Other items include patents, value engineering, safety, documentation standards and product liability.

● RAM Predictions – This focuses attention on the critical failure areas, highlights failures which are difficult to diagnose and provides a measure of the design reliability against the objectives. FMEA, FTA and other modelling exercises are used, in the design reviews, to measure conformance to the RAM targets.
● Design Trade-Offs – These may be between R and M and may involve sacrificing one for the other as, for example, between the reliability of the wrapped joint and the easy replaceability of a connector. Major trade-offs will involve the design review whereas others will be made by the designer.
● Prototype Tests – These cover marginal, functional, parametric, environmental and reliability tests. It is the first opportunity to observe reliability in practice and to make some comparison against the predictions.
● Parts Selection and Approval – Involves field tests or seeking field information from other users. The continued availability of each part is important and may influence the choice of supplier.
● Demonstrations – Since these involve statistical sampling, test plans have to be calculated at an early stage so that the risks can be evaluated.
● Spares Provisioning – This affects reliability and maintainability and has to be calculated during design.
● Data Collection and Failure Analysis – Failure data, with the associated stress information, is essential to reliability growth programmes and also to future predictions. A formal failure-reporting scheme should be set up at an early stage so that tests on the earliest prototype modules contribute towards the analysis.
● Reliability growth – Establishing reporting and analysis to confirm that field reliability growth meets targets.
● Training – Design engineers should be trained to a level where they can work with the R and M specialist. Customer training of maintenance staff is another aspect which may arise.

Figure 18.2

18.4 RESPONSIBILITIES

RAM is an integral part of the design process. In many cases mere lip service is given to it and this leads to little more than high level predictions being carried out too late in the design. These have no effect whatever in bringing the design nearer to the targets. Reliability and maintainability are engineering parameters and the responsibility for their achievement is therefore primarily with the design team. Quality assurance techniques play a vital role in achieving the goals but cannot be used to 'test in' reliability to a design which has its own inherent level. Three distinct responsibilities therefore emerge which are complementary but do not replace each other. See Figure 18.2.

18.5 STANDARDS AND GUIDANCE DOCUMENTS

There are a number of standards which might be called for. The more important are as follows:

- BS 5760: Reliability of systems, equipment and components: This is in a number of parts. Part 1 is Guide to Reliability Programme Management and outlines the reliability activities such as have been dealt with in this book. Other parts deal with prediction, data, practices and so on.
- UK Ministry of Defence Standard 00-40 Reliability and maintainability: This is in eight parts. Parts 1 and 2 are concerned with project requirements and the remainder with requirements documents, training, procurement and so on.
- US Military Standard 785A Reliability Program for Systems and Equipment Development and Production: Specifies programme plans, reviews, predictions and so on.
- US Military Standard 470 Maintainability Programme Requirements: A document, from 1966, which covers the programme plan and specifies activities for design criteria, design review, trade-offs, data collection, predictions and status reporting.

19 Contract clauses and their pitfalls

19.1 ESSENTIAL AREAS

Since the late 1950s in the United States, reliability and maintainability requirements have appeared in both military and civil engineering contracts. These contracts often carry penalties for failure to meet these objectives. For 30 years in the UK, suppliers of military and commercial electronic and telecommuncation equipment have also found that clauses specifying reliability and maintainability were being included in invitations to tender and in the subsequent contracts. Suppliers of highly reliable and maintainable equipment are often well able to satisfy such conditions with little or no additional design or manufacturing effort, but incur difficulty and expense since a formal demonstration of these parameters may not have been previously attempted. Furthermore, a failure-reporting procedure may not exist and therefore historical data as to a product's reliability or repair time may be unobtainable.

The inclusion of system-effectiveness parameters in a contract involves both the suppliers of good and poor equipment in additional activities. System Effectiveness clauses in contracts range from a few words – specifying availability, failure rate or MTBF of all or part of the system – to many pages containing details of design and test procedures, methods of collecting failure data, methods of demonstrating reliability and repair time, limitations on component sources, limits to size and cost of test equipment, and so on. Two types of pitfall arise from such contractual conditions:

1. Those due to the omission of essential conditions or definitions;
2. Those due to inadequately worded conditions which present ambiguities, concealed risks, eventualities unforeseen by both parties, etc.

The following headings are essential if reliability or maintainability is to be specified.

19.1.1 Definitions

If a mean time to repair or down time is specified, then the meaning of repair time must be defined in detail. Mean time to repair is often used when it is mean down time which is intended.

Failure itself must also be thoroughly defined at system and module levels. It may be necessary to define more than one type of failure (for example, total system failure or degradation failure) or failures for different operating modes (for example, in flight or on ground) in order to describe all the requirements. MTBFs might then be ascribed to the different failure types. MTBFs and failure rates often require clarification as to the meaning of 'failure' and 'time'. The latter may refer to operating time, revenue time, clock time, etc. Types of failure

Figure 19.1

which do not count for the purpose of proving the reliability (for example, maintenance induced or environment outside limits) have also to be defined.

For process-related equipment it is usual to specify Availability. Unless, however, some failure modes are defined, the figures can be of little value. For example, in a safety system, failure may consist of spurious alarm or of failure to respond. Combining the two failure rates produces a misleading figure and the two modes must be evaluated separately. Figure 19.1 reminds us of the Bathtub Curve with early, random and wearout failures. Reliability parameters usually refer to random failures unless stated to the contrary, it being assumed that burn-in failures are removed by screening and wearout failures eliminated by preventive replacement.

It should be remembered that this is a statistical picture of the situation and that, in practice, it is rarely possible to ascribe a particular failure to any of these categories. It is therefore vital that, if reliability is being demonstrated by a test or in the field, these early and wearout failures are eliminated, as far as possible, by the measures already described. The specification should make clear which types of failure are being observed in a test.

Parameters should not be used without due regard to their meaning and applicability. Failure rate, for example, has little meaning except when describing random failures. Remember that in systems involving redundancy, constant failure rate may not apply except in the special cases outlined in Chapters 7 to 9. Availability, MTBF or reliability should then be specified in preference.

Reliability and maintainability are often combined by specifying the useful parameter, Availability. This can be defined in more than one way and should therefore be specifically defined. The usual form is the Steady State Availability, which is MTBF/(MTBF + MDT), where MDT is the Mean Down Time.

19.1.2 Environment

A common mistake is to fail to specify the environmental conditions under which the product is to work. The specification is often confined to temperature range and maximum humidity, and this is not always adequate. Even these two parameters can create problems, as with temperature cycling under high-humidity conditions. Other stress parameters include pressure, vibration and shock, chemical and bacteriological attack, power supply variations and interference, radiation, human factors and many others. The combination or the cycling of any of these parameters can have significant results.

Where equipment is used as standby units or held as spares, the environmental conditions will be different to those experienced by operating units. It is often assumed that because a unit is not powered, or in store, it will not fail. In fact the environment may be more conducive to failure under these circumstances. Self-generated heat and mechanical self-cleaning wiping actions are often important ingredients for reliability. If equipment is to be transported while the supplier is liable for failure, then the environmental conditions must be evaluated. On the other hand, over-specifying environmental conditions is a temptation which leads to over-design and higher costs. Environmental testing is expensive, particularly if large items of equipment are involved and if vibration tests are called for. These costs should be quantified by obtaining quotations from a number of test houses before any commitment is made to demonstrate equipment under environmental conditions.

Maintainability can also be influenced by environment. Conditions relating to safety, comfort, health and ergonomic efficiency will influence repair times since the use of protective clothing, remote-handling devices, safety precautions, etc. increases the active elements of repair time by slowing down the technician.

19.1.3 Maintenance support

The provision of spares, test equipment, personnel, transport and the maintenance of both spares and test equipment is a responsibility which may be divided between supplier and customer or fall entirely on either. These responsibilities must be described in the contract and the supplier must be conscious of the risks involved in the customer not meeting his or her side of the bargain.

If the supplier is responsible for training the customer's maintenance staff then levels of skill and training have to be laid down.

Maintenance philosophy, usually under customer control, plays a part in determining reliability. Periodic inspection of a non-attended system during which failed redundant units are changed yields a different MTBF to the case of immediate repair of failed units, irrespective of whether they result in system failure. The maintenance philosophy must therefore be defined.

A contract may specify an MTTR supported by a statement such as 'identification of faulty modules will be automatic and will be achieved by automatic test means. No additional test equipment will be required for diagnosis'. This type of requirement involves considerable additional design effort in order to permit all necessary diagnostic signals to be made accessible and for measurements to be made. Additional hardware will be required either in the form of BITE or an 'intelligent' portable terminal with diagnostic capability. If such a requirement is overlooked when costing and planning the design the subsequent engineering delay and cost is likely to be considerable.

19.1.4 Demonstration and prediction

The supplier might be called upon to give a statistical demonstration of either reliability or repair time. In the case of maintainability a number of corrective or preventive maintenance actions will be carried out and a given MTTR (or better) will have to be achieved for some proportion of the attempts. In this situation it is essential to define the tools and equipment to be used, the maintenance instructions, test environment and technician level. The method of task selection, the spares and the level of repair to be carried out also require stating. The probability of failing the test should be evaluated since some standard tests carry high supplier's risks. When reliability is being demonstrated then a given number of hours will be accumulated and a number of failures stated, above which the test is failed. Again, statistical risks apply and the supplier needs to calculate the probability of failing the test with good equipment and the customer that of passing inadequate goods.

Essential parameters to define here are environmental conditions, allowable failures (for example, maintenance induced), operating mode, preventive maintenance, burn-in, testing costs. It is often not possible to construct a reliability demonstration which combines sensible risks ($\leq 15\%$) for both parties with a reasonable length of test. Under these circumstances the acceptance of reliability may have to be on the basis of accumulated operating hours on previously installed similar systems.

An alternative to statistical or historical demonstrations of repair time and reliability is a guarantee period wherein all or part of the failure costs, and sometimes redesign costs, are borne by the supplier. In these cases great care must be taken to calculate the likely costs. It must be remembered that if 100 items of equipment meet their stated MTBF under random failure conditions, then after operating for a period equal to one MTBF, 63 of them, on average, will have failed.

From the point of view of producer's risk, a warranty period is a form of reliability demonstration since, having calculated the expected number of failures during the warranty, there is a probability that more will occur. Many profit margins have been absorbed by the unbudgeted penalty maintenance arising from this fact.

A prediction is often called for as a type of demonstration. It is desirable that the data source is agreed between the two parties or else the Numbers Game will ensue as various failure rates are 'negotiated' by each party seeking to turn the prediction to his or her favour.

19.1.5 Liability

The exact nature of the supplier's liability must be spelt out, including the maximum penalty which can be incurred. If some qualifying or guarantee period is involved it is necessary to define when this commences and when the supplier is free of liability. The borders between delivery, installation, commissioning and operation are often blurred and therefore the beginning of the guarantee period will be unclear.

It is wise to establish a mutually acceptable means of arbitration in case the interpretation of later events becomes the subject of a dispute. If part of the liability for failure or repair is to fall on some other contractor, care must be taken in defining each party's area. The interface between equipment guaranteed by different suppliers may be physically easy to define but there exists the possibility of failures induced in one item of equipment owing to failure or degraded performance in another. This point should be considered where more than one supplier is involved.

19.2 OTHER AREAS

The following items are often covered in a detailed invitation to tender.

19.2.1 Reliability and maintainability programme

The detailed activities during design, manufacturing and installation are sometimes spelt out contractually. In a development contract this enables the customer to monitor the reliability and maintainability design activities and to measure progress against agreed milestones. Sometimes standard programme requirements are used as, for example:

US Military Standard 470, Maintainability Program Requirements.
US Military Standard 785, Requirements for Reliability Program.

BS 4200: Part 5 Reliability programmes for equipment.
BS 5760 Reliability of constructed and manufactured products, systems, equipment and components.

Typical activities specified are:

Prediction – Data sources, mathematical models.
Testing – Methods and scheduling of design, environmental and other tests.
Design Review – Details of participation in design reviews.
Failure Mode and Effect Analysis – Details of method and timing.
Failure Reporting – Failure reporting documents and reporting procedures.

19.2.2 Reliability and maintainability analysis

The supplier may be required to offer a detailed reliability or maintainability prediction together with an explanation of the techniques and data used. Alternatively, a prediction may be requested using defined data and methods of calculation. Insistence on optimistic data makes it more difficult to achieve the predicted values whereas pessimistic data leads to over-design.

19.2.3 Storage

The equipment may be received by the customer and stored for some time before it is used under conditions different to normal operation. If there is a guarantee period then the storage conditions and durations will have to be defined. The same applies to storage and transport of spares and test equipment.

19.2.4 Design standards

Specific design standards are sometimes described or referenced in contracts or their associated specifications. These can cover many areas, including:

Printed-board assemblies – design and manufacture
Wiring and soldering
Nuts, bolts and threads
Finishes
Component ratings
Packaging

A problem exists that these standards are very detailed and most manufacturers have their own version. Although differences exist in the fine detail they are usually overlooked until some formal acceptance inspection takes place, by which time retrospective action is difficult, time consuming and costly.

19.3 PITFALLS

The foregoing lists those aspects of reliability and maintainability likely to be mentioned in an invitation to tender or in a contract. There are pitfalls associated with the omission or inadequate definition of these factors and some of the more serious are outlined below.

19.3.1 Definitions

The most likely area of dispute is the definition of what constitutes a failure and whether or not a particular incident ranks as one or not. There are levels of failure (system, unit), types of failure (catastrophic, degradation), causes of failure (random, systematic, over-stress) and there are effects of failure (dormant, hazardous). For various combinations of these, different MTBF and MTTR objectives with different penalties may be set. It is seldom sufficient, therefore, to define failure as not performing to specification since there are so many combinations covered by that statement. Careful definition of the failure types covered by the contract is therefore important.

19.3.2 Repair time

It was shown in Chapter 2 that repair times could be divided into elements. Initially they can be grouped into active and passive elements and, broadly speaking, the active elements are dictated by system design and the passive by maintenance and operating arrangements. For this reason, the supplier should never guarantee any part of the repair time which is influenced by the user.

19.3.3 Statistical risks

A statistical maintainability test is described by a number of repair actions and an objective MTTR which must not be exceeded on more than a given number of attempts. A reliability test involves a number of hours and a similar pass criterion of a given number of failures. In both cases producer and consumer risks apply, as explained in earlier chapters, and unless these risks are calculated they can prove to be unacceptable. Where published test plans are quoted, it is never a bad thing to recalculate the risks involved. It is not difficult to find a test which requires the supplier to achieve an MTBF 50 times the value which is to be proved in order to stand a reasonable chance of passing the test.

19.3.4 Quoted specifications

Sometimes a reliability or maintainability programme or test plan is specified by calling up a published standard. Definitions are also sometimes dealt with in this way. The danger with blanket definitions lies in the possibility that not all the quoted terms are suitable and that the standards will not be studied in every detail.

19.3.5 Environment

Environmental conditions affect both reliability and repair times. Temperature and humidity are the most usual to be specified and the problem of cycling has already been pointed out. If other factors are likely to be present in field use then they must either be specifically excluded from the range of environment for which the product is guaranteed or included, and therefore allowed for in the design and in the price. It is not desirable to specify every parameter possible, since this leads to over-design.

19.3.6 Liability

When stating the supplier's liability it is important to establish its limit in terms of both cost and time. Suppliers must ensure that they know when they are finally free of liability.

19.3.7 In summary

The biggest pitfall of all is to assume that either party wins any advantage from ambiguity or looseness in the conditions of a contract. In practice, the hours of investigation and negotiation which ensue from a dispute far outweigh any advantage that might have been secured, to say nothing of the loss of goodwill and reputation. If every effort is made to cover all the areas discussed as clearly and simply as possible, then both parties will gain.

19.4 PENALTIES

There are various ways in which a penalty may be imposed on the basis of maintenance costs or the cost of system outage. It must be remembered that any cash penalty must be a genuine and reasonable pre-estimate of the damages thought to result. Some alternatives are briefly outlined.

19.4.1 Apportionment of costs during guarantee

Figure 19.2(a) illustrates the method where the supplier pays the total cost of corrective maintenance during the guarantee period. He or she may also be liable for the cost of redesign made necessary by systematic failures. In some cases the guarantee period recommences for those parts of the equipment affected by modifications. A disadvantage of this arrangement is that it gives the customer no great incentive to minimize maintenance costs until the guarantee has expired. If the maintenance is carried out by the customer and paid for by the supplier then the latter's control over the preventive maintenance effectiveness is minimal. The customer should never be permitted to benefit from poor maintenance, for which reason this method is not very desirable.

An improvement of this is obtained by Figure 19.2(b), whereby the supplier pays a proportion of the costs during the guarantee and both parties therefore have an incentive to minimize costs. In Figure 19.2(c) the supplier's proportion of the costs decreases over the liability period. In Figure 19.2(d) the customer's share of the maintenance costs remains constant and the supplier pays the excess. The arrangements in (b) and (c) both provide mutual incentives. Arrangement (d), however, provides a mixed incentive. The customer has, initially, a very high incentive to reduce maintenance costs but once the ceiling has been reached this disappears. On the other hand, (d) recognizes the fact that for a specified MTBF the customer should anticipate a given amount of repair. Above this amount the supplier pays for the difference between the achieved and contracted values.

19.4.2 Payment according to down time

The above arrangements involve penalties related to the cost of repair. Some contracts, however, demand a payment of some fixed percentage of the contract price during the down time. Providing that the actual sum paid is less than the cost of the repair this method is similar to

Figure 19.2 Methods of planning penalties

Figure 19.2(b), although in practice it is not likely to be so generous. In any case, an arrangement of this type must be subject to an upper limit.

19.4.3 In summary

Except in case (a) it would not be practicable for the supplier to carry out the maintenance. Usually the customer carries out the repairs and the supplier pays according to some agreed rate. In this case the supplier must require some control over the recording of repair effort and a right to inspect the customer's maintenance records and facilities from time to time. It should be remembered that achievement of reliability and repair time objectives does not imply zero maintenance costs. If a desired MTBF of 20 000 h is achieved for each of ten items of equipment, then in one year (8760 h) about four failures can be expected. On this basis (d) is fairer than (a). When part of a system is subcontracted to another supplier, then the prime contractor must ensure that he or she passes on an appropriate allocation of the reliability commitments in order to be protected.

19.5 SUBCONTRACTED RELIABILITY ASSESSMENTS

It is common, in the development of large systems, for either the designer or the customer to subcontract the task of carrying out Failure Mode Analysis and Reliability Predictions. It may be that the customer requires the designer to place such a contract with a firm of consultants approved by the customer. It is desirable for such work to be covered by a contract which outlines the scope of work and the general agreement between the two parties. Topics to be covered include:

Data bank sources to be used;
  Traceability where non-published data are used;
  Target Reliability, Availability or MTBF;
  Specific duty cycles and environmental profiles;
  Extent of the Failure Mode Analysis required;
  Types of recommendation required in the event of the prediction indicating that the design
    will not meet the objectives;
  Requirement for ranking of major contributors to system failure;
  If the prediction indicates that the design more than meets the objective, a requirement to
    identify the areas of over-design;
  Identification of critical single-point or Common Cause failures;
  Identification of Safety Hazards;
  Recommendations for maintenance (e.g. replacement strategy, periodic inspection time);
  Calculations of spares-holding levels for defined probabilities of stockout;
  Aspects of Human Error required in the analysis;
  Arrangements for control and review of the assessment work, including reporting (e.g.
    Conceptual design report, Interim prediction and report, Detailed Failure Mode Analysis,
    Final Design Qualification report, etc.);
  Schedules, Costs, Invoicing.

19.6 EXAMPLE

The following requirements might well be placed in an invitation to tender for a piece of measuring equipment. They are by no means intended as a model contract and the reader might care to examine them from both the designer's and customer's points of view.

Loss of Measurement shall include the total loss of temperature recording as well as a loss of recording accuracy exceeding 20%.

Mode 1: The loss of 2 or more consecutive measurements.
Mode 2: The loss of recording accuracy of temperature within the range ($>1\%$ to 20%).

Bidders shall satisfy 'XYZ' that the equipment will meet the following requirements.

MTBF (Mode 1) $\geq$ 5 years
MTBF (Mode 2) $\geq$ 10 years

The MTBF shall be achieved without the use of redundancy but by the use of appropriate component quality and stress levels. It shall be demonstrated by means of a failure mode analysis of the component parts. FARADIP.THREE shall be used as the failure rate data source except where alternative sources are approved by 'XYZ'.

The above specification takes no account of the infant mortality failures usually characterized by a decreasing failure rate in the early life of the equipment. The supplier shall determine a suitable burn-in period and arrange for the removal of these failures by an appropriate soak test.

No wearout failure mechanisms, characterized by an increasing failure rate, shall be evident in the life of the equipment. Any components requiring preventive replacement in order to achieve this requirement shall be highlighted to 'XYZ' for consideration and approval.

In the event of the MTBFs not being demonstrated, at 80% confidence, after 10 device years of operation have been accumulated then the supplier will carry out any necessary redesign and modification in order to achieve the MTBF objectives.

During the life of the equipment any systematic failures shall be dealt with by the supplier, who will carry out any necessary redesign and modification. A systematic failure is one which occurs three or more times for the same root cause.

20 Product liability and safety legislation

Product liability is the liability of a supplier, designer or manufacturer to the customer for injury or loss resulting from a defect in that product. There are reasons why it has recently become the focus of attention. The first is the publication in July 1985 of a directive by the European Community, and the second is the wave of actions under United States law which has resulted in spectacular awards for claims involving death or injury. By 1984, sums awarded resulting from court proceedings often reached $1 million. Changes in the United Kingdom became inevitable and the Consumer Protection Act reinforces the application of strict liability. It is necessary, therefore, to review the legal position.

20.1 THE GENERAL SITUATION

20.1.1 Contract law

This is largely governed by the Sale of Goods Act 1979, which requires that goods are of merchantable quality and are reasonably fit for the purpose intended. Privity of Contract exists between the buyer and seller which means that only the buyer has any remedy for injury or loss and then only against the seller, although the cascade effect of each party suing, in turn, the other would offset this. However, exclusion clauses are void for consumer contracts. This means that a condition excluding the seller from liability would be void in law. Note that a contract does not have to be in writing and that a sale, in this context, implies the existence of a contract.

20.1.2 Common law

The relevant area is that relating to the Tort of Negligence, for which a claim for damages can be made. Everyone has a duty of care to his or her neighbour, in law, and failure to exercise reasonable precautions with regard to one's skill, knowledge and the circumstances involved constitutes a breach of that care. A claim for damages for common law negligence is, therefore, open to anyone and not restricted as in Privity of Contract. On the other hand, the onus is with the plaintiff to prove negligence which requires proof:

That the product was defective.
That the defect was the cause of the injury.
That this was foreseeable and that the plaintiff failed in his or her duty of care.

20.1.3 Statute law

The main Acts relevant to this area are:

Sale of Goods Act 1979.
    Goods must be of Merchantable Quality.
    Goods must be fit for purpose.
Unfair Contract Terms Act 1977.
    Exclusion of personal injury liability is void.
    Exclusion of damage liability only if reasonable.
Consumer Protection Act 1987.
    Imposes strict liability.
    Replaces the Consumer Safety Act 1978.
Health and Safety at Work Act 1974, Section 6.
    Involves the criminal law. Places a duty to construct and install items, processes and materials
    without health or safety risks. It applies to places of work. Responsibility involves everyone
    including management. The Consumer Protection Act extends Section 6 of the Health and
    Safety at Work Act to include all areas of use. European legislation will further extend this
    (see Section 20.4.5)

20.1.4 In summary

The present situation involves a form of strict liability but:

Privity of Contract excludes third parties in contract claims.
The onus is to prove negligence unless the loss results from a breach of contract.
Exclusion clauses, involving death and personal injury, are void.

20.2 STRICT LIABILITY

20.2.1 Concept

The concept of strict liability hinges on the idea that liability exists for no other reason than the mere existence of a defect. No breach of contract, or act of negligence, is required in order to incur responsibility and their manufacturers will be liable for compensation if products cause injury.

The various recommendations which are summarized later involve slightly different interpretations of strict liability ranging from the extreme case of everyone in the chain of distribution and design being strictly liable, to the manufacturers being liable unless they can prove that the defect did not exist when the product left them. The Consumer Protection Act makes manufacturers liable whether or not they were negligent.

20.2.2 Defects

A defect for the purposes of product liability, includes:

Manufacturing   – Presence of impurities or foreign bodies.
                 – Fault or failure due to manufacturing or installation.

Design   – Product not fit for the purpose stated.
                        – Inherent safety hazard in the design.
Documentation   – Lack of necessary warnings.
                        – Inadequate or incorrect operating and maintenance instructions result-
                          ing in a hazard.

20.3 THE CONSUMER PROTECTION ACT 1987

20.3.1 Background

In 1985, after 9 years of discussion, the European Community adopted a directive on product liability and member states were required to put this into effect before the end of July 1988. The Consumer Protection Bill resulted in the Consumer Protection Act 1987, which establishes strict liability as described above.

20.3.2 Provisions of the Act

The Act will provide that a producer (and this includes manufactuers, those who import from outside the EC and retailers of 'own brands') will be liable for damage caused wholly or partly by defective products which include goods, components and materials but exclude unprocessed agricultural produce. 'Defective' is defined as not providing such safety as people are generally entitled to expect, taking into account the manner of marketing, instructions for use, the likely uses and the time at which the product was supplied. Death, personal injury and damage (other than to the product) exceeding £275 are included.

The consumer must show that the defect caused the damage but no longer has the onus of proving negligence. Defences include:

● The state of scientific and technical knowledge at the time was such that the producer could not be expected to have discovered the defect. This is known as the 'development risks' defence.
● The defect results from the product complying with the law.
● The producer did not supply the product.
● The defect was not present when the product was supplied by the manufacturer.
● The product was not supplied in the course of business.
● The product was in fact a component part used in the manufacture of a further product and the defect was not due to this component.

In addition, the producer's liability may be reduced by the user's contributory negligence. Further, unlike the privity limitation imposed by contract law, any consumer is covered in addition to the original purchaser.

The Act sets out a general safety requirement for consumer goods and applies it to anyone who supplies goods which are not reasonably safe having regard to the circumstances pertaining. These include published safety standards, the cost of making goods safe and whether or not the goods are new.

20.4 HEALTH AND SAFETY AT WORK ACT 1974

20.4.1 Scope

Section 6 of this Act applies strict liability to articles produced for use at work, although the Consumer Protection Act extends this to all areas. It is very wide and embraces designers, manufacturers, suppliers, hirers and employers of industrial plant and equipment. We are now dealing with criminal law and failure to observe the duties laid down in the Act is punishable by fine or imprisonment. Claims for compensation are still dealt with in civil law.

20.4.2 Duties

The main items are:

To design and construct products without risk to health or safety.
To provide adequate information to the user for safe operation.
To carry out research to discover and eliminate risks.
To make positive tests to evaluate risks and hazards.
To carry out tests to ensure that the product is inherently safe.
To use safe methods of installation.
To use safe (proven) substances and materials.

20.4.3 Concessions

The main concessions are:

- It is a defence that a product has been used without regard to the relevant information supplied by the designer.
- It is a defence that the design was carried out on the basis of a written undertaking by the purchaser to take specified steps sufficient to ensure the safe use of the item.
- One's duty is restricted to matters within one's control.
- One is not required to repeat tests upon which it is reasonable to rely.

20.4.4 Responsibilities

Basically, everyone concerned in the design and provision of an article is responsible for it. Directors and managers are held responsible for the designs and manufactured articles of their companies and are expected to take steps to assure safety in their products. Employees are also responsible. The 'buck' cannot be passed in either direction.

20.4.5 European Community legislation

In 1989/1990 the EC agreed to a framework of directives involving health and safety. This legislation will eventually replace the Health and Safety at Work Act, being more prescriptive and detailed that the former. The directive mirrors the Health and Safety at Work Act by setting general duties on both employees and employers for all work activities.

In implementing this European legislation the Health and Safety Commission will attempt to avoid disrupting the framework which has been established by the Health and Safety at Work Act. The directive covers:

> The overall framework
> The workplace
> Use of work equipment
> Use of personal protective equipment
> Manual handling
> Display screen equipment

20.4.6 Management of Health and Safety at work Regulations 1992

These lay down broad general duties which apply to almost all Great Britain on shore and offshore activities. They are aimed at improving health and safety management and can be seen as a way of making more explicit what is called for by the H&SW Act 1974. They are designed to encourage a more systematic and better organized approach to dealing with health and safety, including the use of risk assessment.

20.5 INSURANCE AND PRODUCT RECALL

20.5.1 The effect of Product Liability trends

- An increase in the number of claims.
- Higher premiums.
- The creation of separate Product Liability Policies.
- Involvement of insurance companies in defining quality and reliability standards and procedures.
- Contracts requiring the designer to insure the customer against genuine and frivolous consumer claims.

20.5.2 Some critical areas

- *All Risks* – This means all risks specified in the policy. Check that your requirements are met by the policy.
- *Comprehensive* – Essentially means the same as the above.
- *Disclosure* – The policy holder is bound to disclose any information relevant to the risk. Failure to do so, whether asked for or not, can invalidate a claim. The test of what should be disclosed is described as 'anything the prudent insurer should know'.
- *Exclusions* – The Unfair Contract Terms Act 1977 does not apply to insurance, so read and negotiate accordingly. For example, defects related to design could be excluded and this would considerably weaken a policy from the product liability standpoint.
- *Prompt notification of claims.*

20.5.3 Areas of cover

Premiums are usually expressed as a percentage of turnover and cover is divided into three areas:

Product Liability – Cover against claims for personal injury or loss.
Product Guarantee – Cover against the expenses of warranty/repair.
Product Recall – Cover against the expenses of recall.

20.5.4 Product recall

A design defect causing a potential hazard to life, health or safety may become evident when a number of products are already in use. It may then become necessary to recall, for replacement or modification, a batch of items, some of which may be spread throughout the chain of distribution and others in use. The recall may vary in the degree of urgency depending on whether the hazard is to life, health or merely reputation. A hazard which could reasonably be thought to endanger life or to create a serious health hazard should be treated by an emergency recall procedure. Where less critical risks involving minor health and safety hazards are discovered a slightly less urgent approach may suffice. A third category, operated at the vendor's discretion, applies to defects causing little or no personal hazard and where only reputation is at risk.

If it becomes necessary to implement a recall the extent will be determined by the nature of the defect. It might involve, in the worst case, every user or perhaps only a specific batch of items. In some cases the modification may be possible in the field and in others physical return of the item will be required. In any case, a full evaluation of the hazard must be made and a report prepared.

One person, usually the Quality Manager, must be responsible for the handling of the recall and must be directly answerable to the Managing Director or Chief Executive. The first task is to prepare, if appropriate, a Hazard Notice in order to warn those likely to be exposed to the risk. Circulation may involve individual customers when traceable, field service staff, distributors, or even the news media. It will contain sufficient information to describe the nature of the hazard and the precautions to be taken. Instructions for returning the defective item can be included, preferably with a pre-paid return card. Small items can be returned with the card whereas large ones, or products to be modified in the field, will be retained while arrangements are made.

Where products are despatched to known customers a comparison of returns with output records will enable a 100% check to be made on the coverage. Where products have been despatched in batches to wholesalers or retail outlets the task is not so easy and the quantity of returns can only be compared with a known output, perhaps by area. Individual users cannot be traced with 100% certainty. Where customers have completed and returned record cards after purchase the effectiveness of the recall is improved.

After the recall exercise has been completed a major investigation into the causes of the defect must be made and the results progressed through the company's Quality and Reliability Programme. Causes could include:

Insufficient test hours
Insufficient test coverage
Insufficient information sought on materials
Insufficient industrial engineering of the product prior to manufacture
Insufficient production testing
Insufficient field/user trials
Insufficient user training

21 Major incident legislation

21.1 HISTORY OF MAJOR INCIDENTS

Since the 1960s, developments in the process industries have resulted in large quantities of noxious and flammable substances being stored and transmitted in locations that could, in the event of failure, affect the public. Society has become increasingly aware of these hazards as a result of major incidents which involve both process plant and public transport such as:

Aberfan (UK)	1966	144 deaths due to collapse of a coalmine waste tip
Flixborough (UK)	1974	28 deaths due to an explosion resulting from the stress failure of a temporary reactor by-pass, leading to an escape of cyclohexane
Beek (Netherlands)	1975	14 deaths due to propylene
Seveso (Italy)	1976	Unknown number of casualties due to a release of dioxin
San Carlos Holiday Camp (Spain)	1978	*c.* 150 deaths due to a propylene tanker accident
Three Mile Island (USA)	1979	0 immediate deaths. Incident due to a complex sequence of operator and physical events following a leaking valve allowing water into the instrument air. This led to eventual loss of cooling and reactor core damage
Bhopal (India)	1984	2000+ deaths following a release of methyl isocyanate due to some safety-related systems being out of service due to inadequate maintenance
Mexico City	1984	500+ deaths due to an LPG explosion at a refinery
Chernobyl (USSR)	1986	31 immediate deaths and unknown number of casualties following the meltdown of a nuclear reactor due to intrinsic reactor design and operating sequences
Herald of Free Enterprise	1987	184 deaths due to capsize of Zeebrugge–Dover ferry
Piper Alpha (North Sea)	1988	167 deaths due to an explosion of leaking condensate following erroneous use of a condensate pump in a stream disenabled for maintenance

Clapham (UK)	1988	34 deaths due to a rail crash resulting from a signalling failure
Kegworth (UK)	1989	47 deaths due to a 737 crash on landing involving erroneous shutdown of the remaining good engine
Cannon Street, London (UK)	1991	2 deaths and 248 injured due to a rail buffer-stop collision
Strasbourg (France)	1992	87 deaths due to A320 Airbus crash
Eastern Turkey	1992	400+ deaths due to methane explosion in a coal mine
Paddington (UK)	1999	31 deaths due to a rail crash (drawing attention to the debate over automatic train protection)
Paris	2000	114 deaths due to the crash of a Concorde aircraft

It is important to note that in a very large number (if not all) of the above incidents human factors played a strong part. It has long been clear that major incidents seldom occur as a result of equipment failure alone but involve humans in the maintenance or operating features of the plant.

Media attention is frequently focused on the effects of such disasters and subsequent inquiries have brought the reasons behind them under increasingly closer scrutiny. The public is now very aware of the risks from major transport and process facilities and, in particular, those arising from nuclear installations. Debate concerning the comparative risks from nuclear and fossil-fuel power generation was once the province of the safety professionals. It is now frequently the subject of public debate. Plant-reliability assessment was, at one time, concerned largely with availability and throughput. Today it focuses equally on the hazardous failure modes.

21.2 DEVELOPMENT OF MAJOR INCIDENT LEGISLATION

Following the Flixborough disaster, in 1974, the Health and Safety Commission set up an Advisory Committee on Major Hazards (ACMH) in order to generate advice on how to handle these major industrial hazards. It made recommendations concerning the compulsory notification of major hazards. Before these recommendations were fully implemented, the Seveso accident, in 1976, drew attention to the lack of formal controls throughout the EC. This prompted a draft European Directive in 1980 which was adopted as the so called Seveso Directive (82/501/EEC) in 1982. Delays in obtaining agreement resulted in this not being implemented until September 1984. Its aim was:

To prevent major chemical industrial accidents and to limit the consequences to people and the environment of any which do occur.

In the UK the HSC (Health and Safety Commission) introduced, in January 1983, the Notification of Installations Handling Hazardous Substances (NIHHS) regulations. These required the notification of hazardous installations and that assessments be carried out of the risks and consequences.

The 1984 EC regulations were implemented, in the UK, as the CIMAH (Control of Industrial Major Accident Hazards regulations, 1984). They are concerned with people and the environment and cover processes and the storage of dangerous substances. A total of 178 substances were listed and the quantities of each which would render them notifiable. In these cases a safety case (nowadays called safety report) is required which must contain a substantial hazard and operability study and a quantitative risk assessment. The purpose of the safety report is to demonstrate either that a particular consequence is relatively minor or that the probability of its occurrence is extremely small. It is also required to describe adequate emergency procedures in the event of an incident. The latest date for the submission of safety reports is 3 months prior to bringing hazardous materials on site.

As a result of lessons learnt from the Bhopal incident there have been two subsequent amendments to the CIMAH regulations (1988 and 1990) which have refined the requirements, added substances and revised some of the notifiable quantities. The first revision reduced the threshold quantities for some substances and the second revision was more comprehensive concerning the storage of dangerous substances.

Following the offshore Piper Alpha incident, in 1988, and the subsequent Cullen enquiry, the responsibility for UK offshore safety was transferred from the Department of Energy to a newly formed department of the HSE (health and Safety Executive). Equivalent requirements to the CIMAH regulations are now applied to offshore installations and the latest date for submitting cases was November 1993.

Quantification of frequency, as well as consequences, in safety reports is now the norm and the role of human error in contributing to failures is attracting increasing interest. Emphasis is also being placed on threats to the environment.

The CIMAH regulations will be replaced by a further directive on the Control of Major Accident Hazards (COMAH). Although similar to CIMAH, the COMAH requirements will be more stringent including:

- Provision of information to the public
- Demonstration of management control systems
- Identification of 'domino' effects
- Details of worker participation

The CIMAH requirements define 'Top Tier' sites by virtue of the threshold quantities of substances. For example, 500 tonnes of bromine, 50 tonnes of acetylene or 100 tonnes of natural gas (methane) render a site 'Top Tier'.

To comply with the top tier regulations a plant operator must:

- Prepare and submit to HSE a safety report
- Draw up an onsite emergency plan
- Provide information to local authorities for an offsite emergency plan
- Provide information to the public
- Report major accidents
- Show, at any time, safe operation

21.3 CIMAH SAFETY REPORTS

The Safety Report provides the HSE with a means of assessing the compliance with the CIMAH regulations. Second, and just as important, the exercise of producing the report increases awareness of the risks and focuses attention on providing adequate protection and mitigation measures. Therefore the safety report must:

- Identify the scale and nature of potential hazards
- Assess the likelihood and consequence of accidents
- Describe the safeguards
- Demonstrate management competence

The contents of a Safety Report are addressed in Schedule 6 of the regulations and include:

- The nature of the dangerous substances, the hazards created by them and the means by which their presence is detected.
- Details of the geography, layout, staffing and processes on the site.
- The procedures for controlling staff and processes (including emergency procedures) in order to provide safe operation.
- A description of the potential accident scenarios and the events and pre-conditions which might lead to them.

QRA (Quantified Risk assessment), whereby frequency as well as the consequences is quantified, is not a specific requirement for onshore Safety Reports. It is, however, becoming more and more the practice to provide such studies as Safety Report support material. For offshore installations QRA is required.

Reports are assessed by the HSE in two stages. The first is a screening process (completed within 6 weeks) which identifies reports clearly deficient in the schedule 6 requirements. Within 12 months a detailed assessment is carried out to reveal any issues which require follow-up action.

A typical Safety report might consist of:

(a) General plant information
    Plant/process description (main features and operating
    conditions)
    Personnel distribution on site
    Local population distribution
(b) Hazard identification
    methodology used
    summary of HAZOP and recommendations
    comparative considerations
    conclusions from hazard identification
(c) Potential hazards and their consequences
    Dangerous materials on site
        Inventory of flammable/dangerous substances
        Hazards created by above
    Analysis and detection of dangerous materials
    Nature of hazards
        Fire and explosion
        Toxic hazards
        Impact/dropped object

        Unloading spillage
        Natural hazards
        Hazards and sources leading to a major accident
(d) Plant Management
        Structure and duties (including responsibilities)
        Personnel qualification
        General manning arrangements
        Operating policy and procedures
        Shift system/transfer of information
        Commissioning and start up of new plant
        Training programmes
        Interface between OM&S Area
        Support functions
        Record keeping
(e) Plant safety features
        Control instrumentation
          Codes and standards
          Integrity
        Electrical distribution
          Design
          Protection
          Changeover
          Recovery
          Emergency generator
          Emergency procedure for power fail
          Isolation for maintenance
          Area classification
        Safety systems
          ESD
          Blowdown
          Relief
        Fire fighting
          Design of system
          Water supplies
          Drenching systems
          Foam
          Halon
          Rendezvous
        Piping design
          Material selection
          Design code
        Plant communications
(f) Emergency Planning
        Onsite emergency plans
        Offsite emergency plan
(g) Other items
        Site meteorological conditions
        Plant and area maps
        Meteorological reports

Health and safety policy
Location of dangerous substances
Site health and safety information sheets
Description of tools used in the analysis

21.4 OFFSHORE SAFETY CASES

The offshore safety case is assessed by the Offshore Safety Division of the HSE and assessment is in two stages:

- an initial screen to determine if the case is suitable for assessment and, if appropriate the preparation of an assessment work plan
- detailed assessment leading to either acceptance or rejection

The content of a safety case needs to cover sufficient detail to demonstrate that:

- the management system is adequate to ensure compliance with statutory health and safety requirements
- adequate arrangements have been made for audit and the preparation of audit reports
- all hazards with the potential to cause a major accident have been identified, their risks evaluated, and measures taken to reduce risks to persons to as low as reasonably practicable

In general, the list of contents shown for CIMAH site safety cases will be suitable. A QRA is obligatory for offshore cases and will include consequences and frequency. Additional items which are specific to offshore are:

- temporary refuge
- control of well pressure
- well and bore details
- seabed properties
- abandonment details

There are three points at which a safety case must be submitted:

- Design
  To be submitted early enough for the detailed design to take account of issues raised.
- Pre-operational
  To be submitted 6 months before operation.
- Abandonment
  To be submitted 6 months before commencement.

Particulars to be covered include:

- *Design safety case for fixed installation*
  Name and address

Safety management system
Scale plan of installation
Scale plan of location, conditions, etc.
Operation and activities
Number of persons on installation
Well operations
Pipelines
Detection equipment
Personnel protection (including performance standards)
QRA
Design and construction codes of practice
Principle features of design and construction

- *Operation safety case for fixed installation*
Name and address
Scale plan of installation
Scale plan of location, conditions, etc.
Operation and activities
Number of persons on installation
Well operations
Pipelines
Detection equipment
Personnel protection (including performance standards)
QRA
Limits of safe operation
Risks are lowest reasonably practicable
Remedial work particulars

- *Safety case for a mobile installation*
Name and address
Scale plan of installation
Operation and activities
Number of persons on installation
Well operations
Detection equipment
Personnel protection (including performance standards)
QRA
Limits of safe operation
Environmental limits
Risks are lowest reasonably practicable
Remedial work particulars

- *Safety case for abandonment of a fixed installation*
Name and address
Scale plan of installation
Scale plan of location, conditions, etc.
Operation and activities

Number of persons on installation
Well operations
Pipelines
Detection equipment
Evacuation details
Wells and pipelines present lowest reasonable risk

21.5 PROBLEM AREAS

Reports must be site specific and the use of generic procedures and justifications is to be discouraged. Adopting the contents of procedures and documents from a similar site is quite valid provided care is taken to ensure that the end result is site specific. Initiating events as well as the impact on surroundings will vary according to the location so it cannot be assumed that procedures adequate for one site will necessarily translate satisfactorily to another. A pressure vessel directly in the flight path of a major airport, or beneath an elevated section of motorway is more at risk from collision than one in a deserted location. A liquid natural gas site on a moor will have different impacts to one situated next to a factory.

The hazards from a dangerous substance may be various and it is necessary to consider secondary as well primary hazards. Natural gas, for example, can asphyxiate as well cause fire and explosion. Furthermore the long term exposure of ground to natural gas will result in the concentration of dangerous trace substances. Decommissioning of gas holder sites therefore involves the removal of such impurities from the soil. Carbon disulphide is hazardous in that it is flammable. However when burned it produces sulphur dioxide which in turn is toxic.

The events which could lead to the major accident scenario have to be identified fully. In other words the fault tree approach (Chapter 8) needs to identify all the initiators of the tree. This is an open ended problem in that it is a subjective judgement as to when they have ALL been listed. An obvious checklist would include, as well as hardware failures:

- Earthquake
- Human error
- Software
- Vandalism/terrorism
- External collision
- Meteorology
- Out of spec substances

The HAZOP approach (Chapter 10) greatly assists in bringing varied views to bear on the problem.

Consequences must also be researched fully. There is a requirement to quantify the magnitude of outcome of the various hazards and the appropriate data and modelling tools are needed. The consequence of a pressurized pipeline rupture, for example, requires the appropriate mathematical treatment for which computer models are available. All eventualities need to be considered such as the meteorological conditions under which the ruptured pipeline will disgorge gas. Damage to the very features which provide protection and mitigation must also be considered when quantifying consequences.

21.6 THE COMAH DIRECTIVE (1999)

The COMAH directive, mentioned above, now replaces CIMAH. It places more emphasis on risk assessment and the main features are:

- The simplification that their application will be dependent on exceeding threshold quantities and the distinction between process and storage will no longer apply.
- The exclusion of explosive, chemical and waste disposal hazards at nuclear installations will be removed. The regulations will not, however, apply to offshore installations.
- Substances hazardous to the environment (as well as people) will be introduced. In the first instance these will take account of the aquatic environment.
- More generic categories of substances will be introduced. The 178 substances currently named will thus reduce to 37. A spin-off is that new substances are more easily catered for by virtue of their inclusion in a generic group.
  More information than before will be publicly available including off site emergency plans.
- The HSE (the competent authority in the UK) will positively approve or reject a safety report.
- The periodic update will be more frequent – 3 years instead of 5 years.
- More onus on demonstrating the effectiveness of proposed safety measures and on showing ALARP.

A key feature of the new regulations is that they cover both safety and the environment. They will be enforced by a competent authority comprising the HSE and the environment agency in England and Wales and the HSE and the Scottish Environmental Protection Agency in Scotland.

21.7 RAIL

The importance of Rail Safety cases was re-enforced, by the Cullen Report, following the public enquiry into the Paddington rail disaster of 1999. The Railway Safety (Miscellaneous Amendments) Regulations 2001 amend the Railway (Safety Case) Regulations 2000 to clarify that train operators' safety cases must cover escape arrangements following emergencies.

22 Integrity of safety-related systems

22.1 SAFETY-RELATED OR SAFETY-CRITICAL?

As well as there being a focus of interest on major accident hazards there is a growing awareness that many failures relate to the control and safety systems used for plant operation and protection. Examples of this type of equipment are Fire Detection Systems, Emergency Shutdown Systems, Distributed Control Systems, Rail Signalling, Automotive Controls, Medical Electronics, Nuclear Control Systems and Aircraft Flight Controls.

Terms such as 'safety-related' and 'safety-critical' have become part of the engineering vocabulary. The distinction between them has become blurred and they have tended to be used synonymously.

'Safety-critical' has tended to be used where the hazard leads to fatality whereas 'Safety-related' has been used in a broader context. There are many definitions all of which differ slightly as, for example:

- some distinguish between multiple and single deaths
- some include injury, illness and incapacity without death
- some include effects on the environment
- some include system damage

However, the current consensus distinguishes them as follows:

- Safety-related systems are those which, singly or together with other safety-related systems, achieve or maintain a safe state for equipment under their control.
- Safety-critical systems are those which, on their own, achieve or maintain a safe state for equipment under their control.

The difference involves the number of levels of protection the term *Safety-Related Application* implies a control or safety function where failure or failures could lead to death, injury or environmental damage.

The term *Safety-Related* applies to any hardwired or programmable system where a failure, singly or in combination with other failures/errors, could lead to death, injury or environmental damage.

A piece of equipment, or software, cannot be excluded from this safety-related category merely by identifying that there are alternative means of protection. This would be to pre-judge the issue and a formal safety integrity assessment would be required to determine the issue.

A distinction is made between control and protection systems. Control systems cause a process to perform in a particular manner whereas protection systems deal with fault conditions and their function is therefore to override the control system. Sometimes the equipment which provides these functions is combined and sometimes it is separate. Both can be safety-related and the relevant issue is whether or not the failure of a particular system can lead to a hazard rather than whether or not it is called a safety system. The argument is often put forward (wrongly) that a system is not safety-related because, in the event of its failure, another level of protection exists. An example might be a circuit for automatically closing a valve in the event of high pressure in a pipeline. This potentially dangerous pressure might also be catered for by the additional protection afforded by a relief valve. This does not, however, mean that the valve closing circuit ceases to be safety-related. It might be the case but this depends upon the failure rates of both the systems and on the integrity target and this will be dealt with shortly.

Until recently the approach has generally been to ensure that, for each possible hazardous failure, there are at least two levels of protection. In other words two independent failures would be necessary in order for the hazard to occur. Using the approach described in the next section a single (simplex) arrangement could be deemed adequate although usually this type of redundancy proves to be necessary in order to make the incident frequencies sufficiently low as to be acceptable.

22.2 SAFETY-INTEGRITY LEVELS (SILs)

Safety-integrity is sometimes defined as the probability of a safety-related system performing the required safety functions under all the stated conditions within a stated period of time. The question arises as to how this may be expressed in terms of a target against which systems can be assessed.

The IEC International Standard 61508, as well as the majority of other standards and guidance (see Section 22.4), adopts the concept of safety-integrity levels. The approach involves setting a SIL target and then meeting both quantitative and qualitative requirements appropriate to the SIL. The higher the SIL the more onerous are the requirements. Table 22.1 shows target figures for four safety-integrity levels of which level 1 is the lowest and level 4 is the highest. The reason for there being effectively two tables (high and low demand) is that there are two ways in which the integrity target may need to be described. The difference can best be understood by way of examples.

Consider the motor car air bag. This is a low demand protection system in the sense that demands on it are infrequent (years or tens of years apart). Thus, the failure rate is of very little use to describe its integrity since failures are dormant and we have also to consider the proof-test interval. What is of interest therefore is the combination of failure rate and down time and we

Table 22.1 Safety-integrity levels

Safety integrity level	High demand rate (Dangerous failures/yr)	Low demand rate (Probability of failure on demand)
4	$\geq 10^{-5}$ to $< 10^{-4}$	$\geq 10^{-5}$ to $< 10^{-4}$
3	$\geq 10^{-4}$ to $< 10^{-3}$	$\geq 10^{-4}$ to $< 10^{-3}$
2	$\geq 10^{-3}$ to $< 10^{-2}$	$\geq 10^{-3}$ to $< 10^{-2}$
1	$\geq 10^{-2}$ to $< 10^{-1}$	$\geq 10^{-2}$ to $< 10^{-1}$

thus specify the probability of failure on demand (PFD). Hence the right hand column of Table 22.1.

On the other hand consider the motor car brakes. Now it is the PFD which is of little use to us because we are using them every few seconds. It is failure rate which is of concern because that is also the rate at which we suffer the hazard. Hence the middle column of Table 22.1.

As an example of selecting an appropriate SIL, assume an involuntary risk scenario (e.g. customer killed by explosion) is accepted as 10^{-5} pa (A). Assume that 10^{-1} of the hazardous events in question lead to fatality (B). Thus the failure rate for the hazardous event can be C = A/B = 10^{-4} pa. Assume a fault tree analysis indicates that the unprotected process only achieves a failure rate of 0.5×10^{-1} pa (D). The failure ON DEMAND of the safety system would need to be E = C/D = 2×10^{-3}. Consulting the right hand column of the table SIL LEVEL 2 is applicable.

There is an alternative approach to establishing Safety integrity levels, known as the Risk Graph approach. This avoids quantifying the maximum tolerable risk of fatality by using qualitative judgements. Figure 22.1 gives an example of a risk graph as used in the UKOOA guidelines (Section 22.4.3).

The advantage is that the graph is easier and quicker to apply but, on the other hand, it is less precise. Interpretations of terms such as 'rare', 'possible', etc. can vary between assessors. There is also the need to calibrate the table in the first place and this is not easy without quantification since the SILs are defined in numerical terms.

Both quantitative and qualitative requirements

It is now necessary to recognize that not all failures can be quantified by use of a failure rate. Random hardware failures tend to be those for which failure rates are available since they relate in general to specific component failures and exhibit, more or less, constant failure rate. On the other hand there are systematic failures, in particular the software related failures dealt with in Chapter 17, for which failure rate prediction is not an option.

Consequence severity	Personnel exposure	Alternatives to avoid danger	Demand rate		
			Relatively high	Low	Very low
Slight injury			–	–	–
	Rare	Possible	1	–	–
Serious injury or 1 death		Not likely	2	1	–
	Frequent	Possible	2	1	1
		Not likely	3	2	1
Multiple deaths	Rare		3	3	2
	Frequent		NR	3	3
Catastrophic			NR	NR	NR

– = No special safety features required
NR = Not recommended. Consider alternatives

Safety Integrity Level (SIL)

Figure 22.1 Risk graph

Therefore, both qualitative and quantitative separate assessments are required by the standards. The numerical requirements of Table 22.1 are thus the targets for the reliability prediction of the random hardware failures and the techniques of Chapters 7 to 9 would apply. Also, however, the standards impose qualitative requirements for each SIL, covering the design cycle activities. They include requirements such as were described in Chapter 17:

> Requirements specification
> Design techniques and their documentation
> Review methods and records
> Document and media control
> Programming language
> Test
> Competence
> Fault tolerance
> Commissioning and de-commissioning
> Modifications

A not infrequent misunderstanding is to assume that, if the qualitative requirements of a particular SIL are observed then, the numerical failure targets given in Table 22.1 will automatically be achieved. This is most certainly not the case since the two issues are quite separate. The quantitative targets refer to random hardware failures and are dealt with in Chapters 7 to 9. The qualitative requirements refer to quite different failures whose frequency is NOT quantified.

Functional safety, and safety-integrity levels, are covered in detail in Functional Safety, A Straightforward Guide to IEC 61508, 2001, Smith and Simpson, Butterworth-Heinemann, ISBN 07506 5270 5.

22.3 PROGRAMMABLE ELECTRONIC SYSTEMS (PESs)

PESs are now the most common form of control or safety system although hardwired systems are still sometimes favoured due to their greater visibility in terms of quantified reliability prediction. There has been controversy, since the early 1980s, concerning the integrity of programmable safety-related systems and, as a result, even now non-programmable controls are still widely used.

For many years there was a general principle that no single software error may lead to a hazardous failure. In practice this meant that where programmable control and protection equipment was used a hard-wired or even mechanical/pneumatic protection arrangement was also provided. In this way no software error can cause the hazard without a simultaneous non-programmable failure. At one time integrity studies concentrated on establishing the existence of this arrangement.

With the emergence of the SIL principle the use of a simplex software based safety system has become acknowledged as credible, at the lower SIL levels, provided that it can be demonstrated that the design meets the requirements of the SIL.

There are three basic configurations of system:

- A simplex PES acting alone
- One or more PESs acting in combination with one or more non-programmable systems (including safety monitors)
- A number of PESs acting in combination (with or without diversity)

22.3.1 Simplex PES

Simplex software based systems (with no other form of protection) are often used where previously a simplex non-programmable system has been used. This is sometimes considered adequate, after appropriate risk assessment, for the protection required.

In these cases particular attention needs to be given to self-test in order to reveal dormant failures. The system should also be designed to fail to a safe state.

22.3.2 One or more PES acting in combination with one or more non-programmable features (including safety monitors)

This configuration potentially offers a high level of protection because some, or all, of the functions are duplicated by alternative levels of protection, such as electrical, pneumatic or other means.

A typical example, for ESD (Emergency Shutdown Systems), is to have a duplex PES with mechanical protection devices plus manual push buttons connected directly to remove power from all output circuits.

One application of this override principle is known as the SAFETY MONITOR. It consists of a device which interrogates the safety related parameters (e.g. inputs and outputs) and overrides the action of the PES when the parameters fail to conform to predetermined rules. Figure 22.2 illustrates this arrangement. It is important that the safety monitor itself is designed so that failures within itself lead to a safe condition.

Figure 22.2 Safety monitor

It should not be assumed, however, that safety monitors always fail to a safe condition. Therefore failure mode and effect analysis (FMEA) during the design, followed by suitable testing, should be undertaken. Safety monitors can be hardware devices but are also implemented by means of look-up tables in read only memory (ROM). In that sense there is an element of software in the safety monitor but there is no execution of programmed instructions since no CPU is involved. A typical application would be in burner control.

22.3.3 A number of PESs acting in combination (with or without diversity)

This arrangement of duplicated or triplicated PESs is often used in order to benefit from the increased reliability offered by redundancy. However, if the replicated systems are identical, or even similar, then common cause failures (Section 8.2) can dominate the assessment.

Duplication of hardware, containing identical software, is no defence against the software failures. Redundancy only protects against random hardware failures. If the software in each of the parallel units is identical then software failures are guaranteed to be common mode. Only the use of diverse software can introduce some defence but this must not be thought of as a panacea. Diversity involves more than just separately coded pieces of software

The use of diversity, as a form of software 'redundancy', is a controversial technique. First, it is inadequate to implement diversity using merely separate designers producing separate source code. The commonality of techniques, training and software engineering culture, together with the fact that errors may be created high in the design cycle, means that failures will be propagated throughout both sets of code.

As a minimum requirement, different processors should be used in each element of the redundancy and the separation of effort should occur far higher than the coding stage. Although this would enhance the safety integrity of the design it will never be a total defence since ambiguities and errors at the requirements stage would still propagate through both of the diverse designs. Minimum requirements for diversity are:

- Separate module specifications
- Separate source code
- Different types of CPU
- Separate power sources

Other, desirable requirements would add:

- CPUs with different clock rates
- Different languages
- Different memory types
- Different supply voltages

Software diversity is also referred to as N-version programming, dissimilar software and multi-version software.

22.4 CURRENT GUIDANCE

22.4.1 IEC International Standard 61508: Functional safety – safety related systems – 7 parts

The International Electrotechnical Commission working group documents IEC/SC65A/WG9/45 and IEC/SC65A/WG10/45 addressed software and system safety respectively. These have now become the IEC International Standard which is in 7 parts:

Part 1 General requirements
Part 2 Requirements for electrical/electronic/programmable electronic systems
Part 3 Software requirements
Part 4 Definitions and abbreviations of terms
Part 5 Guidelines on the application of part 1
Part 6 Guidelines on the application of Parts 2 and 3
Part 7 Bibliography of techniques and measures

The first 3 parts are normative (in other words they are the actual standard) and parts 4 to 7 are informative (they provide guidance on the first 3). The standard deals with the safety life-cycle, establishing risk levels, risk reduction measures, hardware reliability and software quality techniques. Safety-integrity levels are specified as described earlier.

Part 1 deals with setting SIL targets as described in this chapter. Part 2 covers the assessment of random hardware failures (ie quantitative assessment). Part 3 provides the qualitative requirements for each of the SILs.

It is the intention that IEC 61508 becomes the umbrella standard and that industry groups will continue to develop specific '2nd tier' guidance such as that published by the Institution of Gas Engineers.

22.4.2 IEC International Standard 61511: Functional safety – safety instrumented systems for the process industry sector

IEC 61511, currently being drafted, is intended as the process industry 2nd tier guidance to IEC 61508. It is likely to be in 3 parts and might be published in 2001:

    Part 1: Requirements
    Part 2: Guidance to support the requirements
    Part 3: Hazard and Risk Assessment techniques

22.4.3 UKOOA: Guidelines for Process Control and Safety Systems on Offshore Installations

Currently issue 2 (1999) this United Kingdom Offshore Operators Association guide offers guidance for control and safety systems offshore. The sections cover:

    The role of control systems in hazard management
    Categorization of systems (by hazard and application)
    System design
    Equipment design
    Operation and Maintenance

There is an appendix addressing software in safety related systems. Safety-integrity levels are described in a similar way to the IEC 61508 standard. The setting of SIL targets is approached by a risk graph (reproduced by kind permission of UKOOA in Figure 22.1) rather than a quantitative approach.

22.4.4 UK MOD Interim Defence Standard 00–55: The procurement of safety critical software in defence equipment

This superseded the old MOD 00–16 guide to achievement of quality in software. It is far more stringent and is perhaps one of the most demanding standards in this area.

Whereas the majority of the documents described here are for guidance, 00–55 is a standard and is intended to be mandatory on suppliers of 'safety critical' software to the MOD. It is unlikely that the majority of suppliers are capable of responding to all of its requirements but the intention is that, over a period of time, industry migrates to its full use.

It deals with software rather than the whole system and its major requirements include:

- The non-use of assembler language
- The use of static analysis
- A preference for formal methods
- The use and approval of a safety plan
- The use of a software quality plan
- The use of a validation plan
- An independent safety auditor

22.4.5 UK MOD Interim Defence Standard 00–56: Hazard Analysis and safety classification of the computer and programmable electronic system elements of defence equipment

Whereas 00–55 addresses the software this document encompasses the entire 'safety critical' system. It calls for HAZOPS (Hazard and Operability Studies) to be carried out on systems and subsystems of safety related equipment supplied to the UK MOD. There are tables to assist in the classification and interpretation of risk classes and activities are called for according to their severity. This is a risk graph approach which establishes SIL targets. Responsibility for safety has to be formally defined as well as the management arrangements for its implementation. It is intended that 00–56 harmonizes with RTCA DO–178B/(EUROCAE ED–12B) and that it should be compatible with IEC 61508.

22.4.6 RTCA DO–178B/(EUROCAE ED–12B) – Software considerations in airborne systems and equipment certification

This is a very detailed and thorough standard which is used in civil avionics to provide a basis for certifying software used in aircraft. Drafted by a EUROCAE/RTCA committee, DO–178B was published in 1992 and replaces an earlier version published in 1985. The qualification of software tools, diverse software, formal methods and user-modified software are now included.

It defines five levels of software criticality from A (software which can lead to catastrophic failure) to E (No effect). The standard provides guidance which applies to levels A to D.

The detailed listing of techniques (67 pages) covers:

> *Systems aspects:* including the criticality levels, architecture considerations, user modifiable software.
> *The software life-cycle*
> *Software planning*
> *Development:* including requirements, design, coding and integration.
> *Verification:* including reviews, test and test environments.
> *Configuration management:* including baselines, traceability, changes, archive and retrieval.
> *Software quality*
> *Certification*
> *Life-cycle data:* describes the data requirements at the various stages in the life-cycle.

Each of the software quality processes/techniques described in the standard are then listed (10 pages) and the degree to which they are required is indicated for each of the criticality levels A to D.

22.4.7 Institution of Gas Engineers IGE/SR/15: Programmable equipment in safety-related applications

This is the Gas industry 2nd tier guidance to IEC 61508. It is applicable to oil and gas and process applications.

SR/15 describes both quantitative and risk matrix approaches to establishing target SILs. More specific design guidance is given for pressure and flow control, gas holder control, burner control and process shutdown systems. An amendment, published in 2000, addresses the setting of maximum tolerable risk targets (fatality levels) and also includes a checklist schedule to aid conformity in the rigour of carrying out assessments.

22.4.8 UK MOD Interim Standard 00-58: A guideline for HAZOP studies on systems which include programmable electronic systems

As the title suggests this standard describes the HAZOP process (Section 10.3.1) in the context of identifying potentially hazardous variations from the design intent. Part 1 is the requirements and Part 2 provides more detailed guidance on such items as HAZOP guide words for particular types of system, team roles, recording the study, etc.

22.4.9 Draft European Standard BSEN 51026: Railway applications – The Specification and Demonstration of Dependability, Reliability, Maintainability and Safety (RAMS)

This is the rail industry 2nd tier guidance. Risks are assessed by the 'risk matrix' approach whereby severity, frequency, consequence, etc. are specified by guidewords and an overall 'risk classification' obtained. The guidance is life-cycle based in that requirements are stated through the design and implementation stages of a project. 50128 covers software for railway control and 50129 the hardware.

22.4.10 UK HSE Guidance Document: Programmable Electronic Systems in Safety Related Application

Published in 1987, the 'HSE Guidelines' have very much dominated the assessment of safety related programmable equipment for over a decade. The document is shortly to be withdrawn and the HSE already adopts the IEC international standard 61508 in its place. Nevertheless it had a profound effect on integrity studies and is still worth a mention.

The guidelines were developed from a draft as early as 1984 which, in turn, arose from a booklet *Microprocessors in Industry* published by the HSE in 1981.

Second-tier (less generic) guidance was encouraged by the HSE and the Institution of Gas Engineers, EEMUA (Electrical and Electronic Manufacturers and Users Association) and UKOOA (UK Offshore Operators Association) documents followed in that spirit. It addressed:

● The configuration
● The hardware reliability
● The systematic, including software, safety integrity

22.5 ACCREDITATION AND CONFORMITY OF ASSESSMENT

Following a DTI initiative in 1998/9, a framework is being set up to provide a system of accreditation and certification for safety-integrity (*vis-a-vis* IEC 61508) in much the same way as the certification of ISO 9000 quality management operates. Currently the scheme is being developed and operated by CASS Ltd (Conformity of Safety-related Systems) and will eventually cover the accreditation of both:

1. Capability of organizations to design and operate safety-related systems
2. The assessments of the safety-related products and projects safety assessors

23 A case study: The Datamet Project

This chapter is a case study which has been used by the author, on Reliability and Maintainability Management and contract courses for nearly 20 years. It is not intended to represent any actual company, product or individuals.

The page entitled 'Syndicate Study' suggests a number of areas for thought and discussion. When discussing the contract clauses two syndicates can assume the two rules of producer and customer, respectively. After individual study and discussion the two syndicates can renegotiate the contract under the guidance of the course tutor. This approach has proved both stimulating and effective. It is worth reflecting, when criticizing the contract reliability clauses, that although the case study is fictional the clauses were drawn from actual examples.

23.1 INTRODUCTION

The Communications Division of Electrosystems Ltd has an annual turnover of £15 million. Current year's sales are forecast as follows:

	Line Communications	h.f. Radio	Special Systems
Home sales	£9 600 000	£2 000 000	£ 300 000
Export	£ 900 000	£ 900 000	£1 200 000

Line communications systems include 12 circuit, 4 MHz and 12 MHz multiplex systems. A highly reliable range of h.f. radio products includes ship-to-shore, radio beacons, SOS equipment, etc. Special systems constitute 10% of sales and consist of equipment for transmitting information from oil wells and pipe lines over line systems.

The structure of the Division, which employs 1000 personnel, is shown in Figure 23.1 and that of the Engineering Department in Figure 23.2.

23.2 THE DATAMET CONCEPT

In June 1978 the Marketing Department was investigating the market potential for a meteorological telemetry system (DATAMET) whereby a number of observations at some remote location could be scanned, in sequence, and the information relayed by v.h.f. radio to a terminal station. Each observation is converted to an analogue signal in the range 0–10 V and

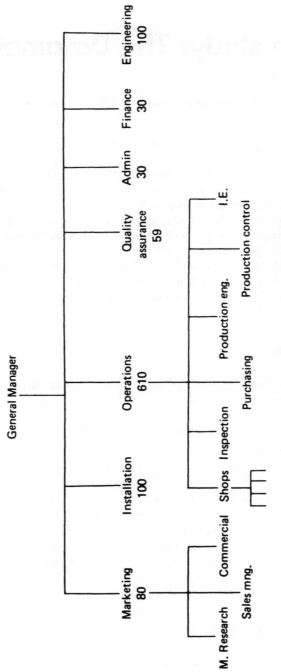

Admin consists of Personnel and Site Services
Numbers indicate approximate numbers of personnel

Figure 23.1

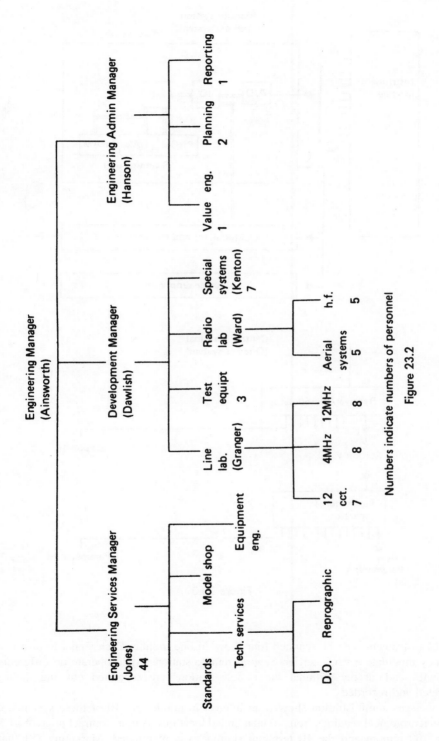

Numbers indicate numbers of personnel

Figure 23.2

Figure 23.3

up to 14 instruments can be scanned four times in one minute. Each signal in turn is used to frequency modulate a v.h.f. carrier. Several remote stations could operate on different carrier frequencies and, at the terminal the remote stations are separated out and their signals interpreted and recorded.

An overseas administration showed an interest in purchasing 10 of these systems, each to relay meteorological readings from 10 unattended locations. A total contract price of £1 500 000 for the 100 remote and the 10 terminal stations was mentioned. Marketing felt that some £6 million of sales could be obtained for these systems over 5 years.

Figure 23.4

23.3 FORMATION OF THE PROJECT GROUP

The original feasibility group consisted of Peter Kenton (Special Systems section head), Len Ward (Radio Lab section head) who had some v.h.f. experience and Arthur Parry (a sales engineer).

A suggested design involved the continuous transmission of each reading on a different frequency. This was found to be a costly solution and, since continuous monitoring was not essential, a scanning system was proposed. Figure 23.3 illustrates the system whereby each instrument reading is converted to an electrical analogue in the 0 - 10 V range. The 14 channels are scanned by a microprocessor controller which sends each signal in code form to the modulator unit. Each remote station operates at a different frequency in the region of 30 MHz. After each cycle of 14 signals a synchronizing signal, outside the normal voltage range, is sent. The terminal station consists of a receiver and demodulator for separating out the remote stations. The signal from each station is then divided into 14 channels and fed to a desktop calculator with printer.

A meteorological equipment supplier was found who was prepared to offer instruments converting each reading to a 0 – 10 V signal. Each set of 14 instruments would cost £1400 for the quantities involved.

Owing to the interest shown by the potential overseas customer it was decided to set up a project group with Kenton as Project manager. The group consisted of Ward and another radio engineer, two special systems engineers, three equipment engineers and four technicians. The project organization, with Kenton reporting to Ainsworth, is shown in Figure 23.4. In September 1978 Kenton prepared the project plan shown in Figure 23.5.

Figure 23.5

23.4 RELIABILITY REQUIREMENTS

In week 5 the customer expressed a firm intention to proceed and the following requirements became known:

Remote stations
MTBF of 5 years
Preventive maintenance at 6-month intervals
Equipment situated in windproof hut with inside temperature range 0 - 50°C
Cost of corrective maintenance for the first year to be borne by supplier

Terminal
MTBF of 2000 h
Maximum repair time of 1 h

(a) Five years mean time between failures is required for each remote station, 2000 h mean time between failures for the terminal. The supplier will satisfy the customer, by means of a reliability prediction, that the design is capable of meeting these objectives.

(b) The equipment must be capable of operating in a temperature range of 0–50°C with a maximum relative humidity of 80%.

(c) Failure shall consist of the loss of any parameter or the incorrect measurement of any parameter.

(d) For one year's operation of the equipment the contractor will refund the cost of all replacements to the terminal equipment and to the remote equipment. When a corrective maintenance visit, other than a routine visit, is required the contractor shall refund all labour and travelling costs including overtime and incentives at a rate to be agreed.

(e) In the event of a system failure then the maximum time to restore the terminal to effective operation shall be 1 h. The contractor is required to show that the design is compatible with this objective.

(f) In the event of systematic failures the contractor shall perform all necessary redesign work and make the necessary modifications to all systems.

(g) The contractor is to use components having the most reasonable chance of being available throughout the life of the equipment and is required to state shelf life and number of spares to be carried in the case of any components that might cease to be available.

(h) The use of interchangeable printed cards may be employed and a positive means of identifying which card is faulty must be provided so that, when a fault occurs, it can be rectified with the minimum effort and skill. The insertion of cards in the wrong position shall be impossible or shall not cause damage to the cards or system.

(i) Maintenance instructions will be provided by the contractor and shall contain all necessary information for the checking and maintenance of the system. These shall be comprehensive and give full operational and functional information. The practice of merely providing a point to point component description of the circuits will not, in itself, be adequate.

Figure 23.6

The first of the 10 systems was to be installed by week 60 and the remainder at 1-month intervals. The penalty maintenance clause was to take effect, for each station, at the completion of installation.

The customer produced a draft contract in week 8 and Parry was asked to evaluate the reliability clauses which are shown in Figure 23.6.

23.5 FIRST DESIGN REVIEW

The first design review was chaired by Ainsworth and took place in week 10. It consisted of Kenton, Parry, Ward, Jones, Ainsworth and the Marketing Manager. Kenton provided the following information:

	Number	λ	$k\lambda$	$Nk\lambda$	
Instruments	14	1.3	–	18.2	
Connections	14	0.0002	0.0003	–	
					18.2

Cyclic Switch					
Microprocessor	1	0.3	0.45	0.45	
Memory chips	3	0.02	0.03	0.09	
MSI chips	2	0.1	0.15	0.3	
Capacitors	15	0.01	0.015	0.225	
Transistors	15	0.05	0.075	1.125	
Solder joints	250	0.0002	0.0003	0.075	
					2.27

Modulator and Transmitter					
Varactors	2	0.1	0.15	0.3	
Transistors	10	0.05	0.075	0.75	
Resistors	30	0.001	0.0015	0.045	
Trimmers	5	0.1	0.15	0.75	
Capacitors	12	0.01	0.015	0.18	
Crystal	2	0.05	0.075	0.15	
Transformer	1	0.2	0.3	0.3	
Solder joints	150	0.0002	0.0003	0.045	
					2.52

Power					
Transformer	1	0.4	0.6	0.6	
Transistors	10	0.1	0.15	1.5	
Zeners	3	0.1	0.15	0.45	
Power diodes	6	0.1	0.15	0.9	
Capacitors (electrolytic)	6	0.1	0.15	0.9	
Solder joints	40	0.0002	0.0003	0.012	
					4.36

$$27.35 \times 10^{-6}$$

$$\underline{\text{Therefore MTBF} = 36\,600\,\text{h} = 4\text{ years}}$$

Figure 23.7

From Figure 23.5 the project group would expend 250 man-weeks.

Engineering assistance would be 70 man-weeks for Drawing, Model Shop, Test equipment building, Technical writing.

All engineering time was costed at £400 per man-week.

The parts for the laboratory model would cost £10 000.

The production model which would involve one terminal and two remote stations would cost £60 000.

Likely production cost for the systems would be £ 100 000 for a terminal with 10 remotes. The above costs did not include the instruments.

On the basis of these costs the project was considered satisfactory if a minimum of four such contracts was to be received.

An initial crude reliability prediction had been carried out by Kenton for the remote equipment and this is reproduced in Figure 23.7. It assumed random failures, generous component tolerancing, commercial components and fixed ground conditions. A multiplication factor of 1.5 was applied to the data to allow for the rather more stringent conditions and a Mean Time Between Failures of about 4 years was obtained. Since no redundancy had been assumed this represented a worst-case estimate and Kenton maintained that the objective of 5 years would eventually be met. Ward, however, felt that the factor of 1.5 was quite inadequate since the available data referred to much more controlled conditions. A factor of 3 would place Kenton's estimate nearly an order below the objective and he therefore held that more attention should be given to reliability at this stage. He was over-ruled by Kenton who was extremely optimistic about the whole project.

The outline design was agreed and it was recorded that attention should be given to:

(a) The LSI devices.
(b) Obtaining an MTBF commitment from the instrument supplier.
(c) Thorough laboratory testing.

23.6 DESIGN AND DEVELOPMENT

The contract, for £1 500 000, was signed in week 12 with two modifications to the reliability section. Parry insisted that the maximum of 1 h for repair should be replaced by a mean time to repair of 30 min since it is impossible to guarantee a maximum repair time. For failures to the actual instruments the labour costs were excluded from the maintenance penalty. Purchasing obtained a 90 years' MTBF commitment from the instrument supplier.

Design continued and by week 20 circuits were being tested and assembled into a laboratory model. Kenton carried out a second reliability prediction in week 21 taking account of some circuit redundancy and of the 6-monthly visits. Ward still maintained that a multiplication factor of 3 was needed and Kenton agreed to a compromise by using 2.5. This yielded an MTBF of 7 years for a remote station. Ward pointed out that even if an MTBF of 8 years was observed in practice then, during the first year, some 12 penalty visits could be anticipated. The cost of a repair involving an unscheduled visit to a remote station could well be in the order of £1200.

At the commencement of laboratory testing Ward produced a failure-reporting format and suggested to Parry that the customer should be approached concerning its use in field reporting. Since a maintenance penalty had been accepted he felt that there should be some control over the customer's failure reporting. In the meantime, the format was used during laboratory testing and Ward was disturbed to note that the majority of failures arose from combinations of drift conditions rather than from catastrophic component failures. Such failures in the field would be likely to be in addition to those anticipated from the predicted MTBF.

In week 30 the supplier of the instruments became bankrupt and it was found that only six sets of instruments could be supplied. With some difficulty, an alternative supplier was found who could provide the necessary instruments. Modifications to the system were required since the new instruments operated over a 0 – 20 V range. The cost was £1600 per set of 14.

23.7 SYNDICATE STUDY

First session
1. Comment on the Project Plan prepared by Kenton.
   (a) What activities were omitted, wrongly timed or inadequately performed?
   (b) How would you have organized this project?
2. Comment on the organization of the project group.
   (a) Do you agree with the reporting levels?
   (b) Were responsibilities correctly assigned?
3. Is this project likely to be profitable? If not in what areas is money likely to be lost?

Second session
1. Discuss the contract clauses and construct alternatives either as
   (i) Producer
   (ii) Customer
2. Set up a role-playing negotiation.

23.8 HINTS

1. Consider the project, and projected figures, as percentage of turnover.
2. Compare the technologies in the proposed design with the established product range and look for differences.
3. Look for the major sources of failure (rate).
4. Consider the instrument reliability requirement and the proposed sourcing.
5. Think about appraisal of the design feasibility.
6. This book has frequently emphasized Allocation.
7. Why is this not a development contract?
8. How were responsibilities apportioned?
9. Were appropriate parameters chosen? (Availability).
10. What were the design objectives?
11. Think about test plans and times.
12. Schedule Design Reviews.
13. Define failure modes and types with associated requirements.

Appendix 1 Glossary

A1 TERMS RELATED TO FAILURE

A1.1 Failure

Termination of the ability of an item to perform its specified function. OR, Non-conformance to some defined performance criteria. Failures may be classified by:

Meaningless without performance spec.

1. Cause –
   Misuse: Caused by operation outside specified stress.
   Primary: Not caused by an earlier failure.
   Secondary: Caused by an earlier failure.
   Wearout: Caused by accelerating failure rate mechanism.
   Design: Caused by an intrinsic weakness.

   Software: Caused by a program error despite no hardware failure
2. Type –
   Sudden: Not anticipated and no prior degradation.
   Degradation: Parametric drift or gradual reduction in performance.
   Intermittent: Alternating between the failed and operating condition.
   Dormant: A component or unit failure which does not cause system failure but which either hastens it or, in combination with another dormant fault, would cause system failure.
   Random: Failure is equally probable in each successive equal time interval.
   Catastrophic: Sudden and complete.

Chapter 2

Chapter 17

A1.2 Failure mode

The outward appearance of a specific failure effect (e.g. open circuit, leak to atmosphere).

Chapter 2

A1.3 Failure mechanism

The physical or chemical process which causes
the failure.

Chapter 11

A1.4 Failure rate

The number of failures of an item per unit time.

Per hour, cycle, opera-
tion, etc.

   This can be applied to:

1. Observed failure rate: as computed from a
   sample.

Point estimate

2. Assessed failure rate: as inferred from sample
   information.

Involves a confidence
level

3. Extrapolated failure rate: projected to other
   stress levels.

A1.5 Mean Time Between Failures and Mean Time to Fail

The total cumulative functioning time of a population divided by the number of failures. As
with failure rate, the same applies to Observed,
Assessed and Extrapolated MTBF. MTBF is used
for items which involve repair. MTTF is used for
items with no repair.

A1.6 Common Cause Failure

The result of an event(s) which, because of
dependencies, causes a coincidence of failure
states of components in two or more separate
channels of a redundant system, leading to the
defined system failing to perform its intended
function.

Section 8.2

A1.7 Common Mode Failure

A subset of *Common Cause Failure* whereby
two or more components fail in the same
manner.

Section 8.2

A2 RELIABILITY TERMS

A2.1 Reliability

The probability that an item will perform a required function, under stated conditions, for a stated period of time.

Since observed reliability is empirical it is defined as the ratio of items which perform their function for the stated period to the total number in the sample.

A2.2 Redundancy

The provision of more than one means of achieving a function.

Active: All items remain operating prior to failure.
Standby: Replicated items do not operate until needed.

A2.3 Diversity

The same performance of a function by two or more independent and dissimilar means (of particular relevance to software).

Chapter 17

A2.4 Failure Mode and Effect Analysis

Determining the outcomes of all known failure modes within an assembly or circuit.

Section 9.3

A2.5 Fault Tree Analysis

A graphical method of modelling a system failure using AND and OR logic in tree form.

Section 8.3

A2.6 Cause Consequence Analysis (Event Trees)

A graphical method of modelling one or more outcomes of a failure or of an event by means of interconnected YES/NO decision boxes.

Section 8.4

A2.7 Reliability growth

Increase in reliability as a result of continued design modifications resulting from field data feedback.

Section 12.3

A2.8 Reliability centred maintenance

The application of quantified reliability techniques to optimizing discard, times, proof-test
intervals and spares levels.

Chapter 16

A3 MAINTAINABILITY TERMS

A3.1 Maintainability

The probability that a failed item will be restored
to operational effectiveness within a given period
of time when the repair action is performed in
accordance with prescribed procedures.

A3.2 Mean Time to Repair (MTTR)

The mean time to carry out a defined maintenance action.

Usually refers to cor-
rective maintenance

A3.3 Repair rate

The reciprocal of MTTR.

When used in reliabil-
ity calculations it is
the reciprocal of Down
Time

A3.4 Repair time

The time during which an item is undergoing
diagnosis, repair, checkout and alignment.

Must be carefully
defined; may also
depend on diagnostics

Chapter 14
and Section
9.2

A3.5 Down Time

The time during which an item is not able to
perform to specification.

Must be carefully
defined

A3.6 Corrective maintenance

The actions associated with repair time.

A3.7 Preventive maintenance

The actions, other than corrective maintenance,
carried out for the purpose of keeping an item in
a specified condition.

A3.8 Least Replaceable Assembly (LRA)

That assembly at which diagnosis ceases and replacement is carried out.

Typically a printed-board assembly

A3.9 Second-line maintenance

Maintenance of LRAs which have been removed from the field for repair or for preventive maintenance.

A4 TERMS ASSOCIATED WITH SOFTWARE

A4.1 Software

All documentation and inputs (for example, tapes, disks) associated with programmable devices.

A4.2 Programmable device

Any piece of equipment containing one or more components which provides a computer architecture with memory facilities.

A4.3 High-level language

A means of writing program instructions using symbols each of which represents several program steps.

A4.4 Assembler

A program for converting program instructions, written in mnemonics, into binary machine code suitable to operate a programmable device.

A4.5 Compiler

A program which, in addition to being an assembler, generates more than one instruction for each statement thereby permitting the use of a high-level language.

A4.6 Diagnostic software

A program containing self-test algorithms enabling failures to be identified.

Particularly applicable to ATE

A4.7 Simulation

The process of representing a unit or system by some means in order to provide some or all identical inputs, at some interface, for test purposes. A means of prediction.

A4.8 Emulation

A type of simulation whereby the simulator responds to all possible inputs as would the real item and generates all the corresponding outputs.

Identical to the real item from the point of view of a unit under test

A4.9 Load test

A system test involving simulated inputs in order to prove that the system will function at full load.

A4.10 Functional test

An empirical test routine designed to exercise an item such that all aspects of the software are brought into use.

A4.11 Software error

An error in the digital state of a system which may propagate to become a failure.

A4.12 Bit error rate

The random incidence of incorrect binary digits.

Expressed 10^{-x}/bit

A4.13 Automatic Test Equipment (ATE)

Equipment for stimulus and measurement controlled by a programmed sequence of steps (usually in software).

A4.14 Data corruption

The introduction of an error by reason of some change to the software already resident in the system. This could arise from electrical interference or from incorrect processing of a portion of the software.

A5 TERMS RELATED TO SAFETY

A5.1 Hazard

A scenario whereby there is a potential for
human, property or environmental damage.

A5.2 Major hazard

A general, imprecise, term for large-scale hazards as, for example, in the chemical or nuclear
industries.

A5.3 Hazard Analysis

A term which refers to a number of techniques   Chapter 10
for analysing the events leading to a hazardous
situation.

A5.4 HAZOP

Hazard and Operability Study – a formal analysis   Chapter 10
of a process or plant by the application of
guidewords.

A5.5 Risk

The likelihood, expressed either as a probability   Chapters 3
or as a frequency, of a hazard materializing.   and 10

A5.6 Consequence analysis

Techniques which involve quantifying the outcome of failures in terms of dispersion, radiation,
fatality etc.

A5.7 Safety-integrity

The probability of a system performing specific
safety functions in a stated period of time.

A5.8 Safety integrity level

One of 4 discrete target levels for specifying
safety integrity requirements.

A6 MISCELLANEOUS TERMS

A6.1 Availability (steady state)

The proportion of time that an item is capable of operating to specification within a large time interval.

Given as: MTBF/(MTBF + MDT)

A6.2 Burn-in

The operation of items for a specified period of time in order to remove early failures and bring the reliability characteristic into the random failure part of the Bathtub Curve.

A6.3 Confidence interval

A range of a given variable within which a random value will lie at a stated confidence (probability).

A6.4 Consumer's risk

The probability of an unacceptable batch being accepted owing to a favourable sample.

A6.5 Derating

The use of components having a higher strength rating in order to reduce failure rate.

A6.6 Ergonomics

The study of human/machine interfaces in order to minimize human errors due to mental or physical fatigue.

A6.7 Mean

Usually used to indicate the Arithmetic Mean, which is the sum of a number of values divided by the number thereof.

A6.8 Median

The median is that value such that 50% of the values in question are greater and 50% less than it.

A6.9 Producer's risk

The probability of an acceptable batch being rejected owing to an unfavourable sample.

A6.10 Quality

Conformance to specification.

A6.11 Random

Such that each item has the same probability of being selected as any other.

A6.12 FRACAS

An acronym meaning failure reporting and corrective action system.

A6.13 RAMS

A general term for reliability, availability, maintainability and safety integrity.

Appendix 2 Percentage points of the Chi-square distribution

α \ n	0·9995	0·999	0·995	0·990	0·975	0·95	0·90	0·80	0·70	0·60
1	0.0^6393	0.0^5157	0.0^4393	0.0^3157	0.0^3982	0.0^2393	0·0158	0·0642	0·148	0·275
2	0.0^2100	0.0^2200	0·0100	0·0201	0·0506	0·103	0·211	0·446	0·713	1·02
3	0·0153	0·0243	0·0717	0·115	0·216	0·352	0·584	1·00	1·42	1·87
4	0·0639	0·0908	0·207	0·297	0·484	0·711	1·06	1·65	2·19	2·75
5	0·158	0·210	0·412	0·554	0·831	1·15	1·61	2·34	3·00	3·66
6	0·299	0·381	0·676	0·872	1·24	1·64	2·20	3·07	3·83	4·57
7	0·485	0·598	0·989	1·24	1·69	2·17	2·83	3·82	4·67	5·49
8	0·710	0·857	1·34	1·65	2·18	2·73	3·49	4·59	5·53	6·42
9	0·972	1·15	1·73	2·09	2·70	3·33	4·17	5·38	6·39	7·36
10	1·26	1·48	2·16	2·56	3·25	3·94	4·87	6·18	7·27	8·30
11	1·59	1·83	2·60	3·05	3·82	4·57	5·58	6·99	8·15	9·24
12	1·93	2·21	3·07	3·57	4·40	5·23	6·30	7·81	9·03	10·2
13	2·31	2·62	3·57	4·11	5·01	5·89	7·04	8·63	9·93	11·1
14	2·70	3·04	4·07	4·66	5·63	6·57	7·79	9·47	10·8	12·1
15	3·11	3·48	4·60	5·23	6·26	7·26	8·55	10·3	11·7	13·0
16	3·54	3·94	5·14	5·81	6·91	7·96	9·31	11·2	12·6	14·0
17	3·98	4·42	5·70	6·41	7·56	8·67	10·1	12·0	13·5	14·9
18	4·44	4·90	6·26	7·01	8·23	9·39	10·9	12·9	14·4	15·9
19	4·91	5·41	6·84	7·63	8·91	10·0	11·7	13·7	15·4	16·9
20	5·40	5·92	7·43	8·26	9·59	10·9	12·4	14·6	16·3	17·8
21	5·90	6·45	8·03	8·90	10·3	11·6	13·2	15·4	17·2	18·8
22	6·40	6·98	8·64	9·54	11·0	12·3	14·0	16·3	18·1	19·7
23	6·92	7·53	9·26	10·2	11·7	13·1	14·8	17·2	19·0	20·7
24	7·45	8·08	9·98	10·9	12·4	13·8	15·7	18·1	19·9	21·7
25	7·99	8·65	10·5	11·5	13·1	14·6	16·5	18·9	20·9	22·6
26	8·54	9·22	11·2	12·2	13·8	15·4	17·3	19·8	21·8	23·6
27	9·09	9·80	11·8	12·9	14·6	16·2	18·1	20·7	22·7	24·5
28	9·66	10·4	12·5	13·6	15·3	16·9	18·9	21·6	23·6	25·5
29	10·2	11·0	13·1	14·3	16·0	17·7	19·8	22·5	24·6	26·5
30	10·8	11·6	13·8	15·0	16·8	18·5	20·6	23·4	25·5	27·4
31	11·4	12·2	14·5	15·7	17·5	19·3	21·4	24·3	26·4	28·4
32	12·0	12·8	15·1	16·4	18·3	20·1	22·3	25·1	27·4	29·4
33	12·6	13·4	15·8	17·1	19·0	20·9	23·1	26·0	28·3	30·3
34	13·2	14·1	16·5	17·8	19·8	21·7	24·0	26·9	29·2	31·3
35	13·8	14·7	17·2	18·5	20·6	22·5	24·8	27·8	30·2	32·3
36	14·4	15·3	17·9	19·2	21·3	23·3	25·6	28·7	31·1	33·3
37	15·0	16·0	18·6	20·0	22·1	24·1	26·5	29·6	32·1	34·2
38	15·6	16·6	19·3	20·7	22·9	24·9	27·3	30·5	33·0	35·2
39	16·3	17·3	20·0	21·4	23·7	25·7	28·2	31·4	33·9	36·2
40	16·9	17·9	20·7	22·2	24·4	26·5	29·1	32·3	34·9	37·1
41	17·5	18·6	21·4	22·9	25·2	27·3	29·9	33·3	35·8	38·1
42	18·2	19·2	22·1	23·7	26·0	28·1	30·8	34·2	36·8	39·1
43	18·8	19·9	22·9	24·4	26·8	29·0	31·6	35·1	37·7	40·0
44	19·5	20·6	23·6	25·1	27·6	29·8	32·5	36·0	38·6	41·0
45	20·1	21·3	24·3	25·9	28·4	30·6	33·4	36·9	39·6	42·0
46	20·8	21·9	25·0	26·7	29·1	31·4	34·2	37·8	40·5	43·0
47	21·5	22·6	25·8	27·4	30·0	32·3	35·1	38·7	41·5	43·9
48	22·1	23·3	26·5	28·2	30·8	33·1	35·9	39·6	42·4	44·9
49	22·8	24·0	27·2	28·9	31·6	33·9	36·8	40·5	43·4	45·9
50	23·5	24·7	28·0	29·7	32·4	34·8	37·7	41·4	44·3	46·9

0·50	0·40	0·30	0·20	0·10		0·05	0·025	0·01	0·005	0·001		0·0005	α / n
0·455	0·708	1·07	1·64	2·71		3·84	5·02	6·63	7·88	10·8		12·1	1
1·39	1·83	2·41	3·22	4·61		5·99	7·38	9·21	10·6	13·8		15·2	2
2·37	2·95	3·67	4·64	6·25		7·81	9·35	11·3	12·8	16·3		17·7	3
3·36	4·04	4·88	5·99	7·78		9·49	11·1	13·3	14·9	18·5		20·0	4
4·35	5·13	6·06	7·29	9·24		11·1	12·8	15·1	16·7	20·5		22·1	5
5·35	6·21	7·23	8·56	10·6		12·6	14·4	16·8	18·5	22·5		24·1	6
6·35	7·28	8·38	9·80	12·0		14·1	16·0	18·5	20·3	24·3		26·0	7
7·34	8·35	9·52	11·0	13·4		15·5	17·5	20·1	22·0	26·1		27·9	8
8·34	9·41	10·7	12·2	14·7		16·9	19·0	21·7	23·6	27·9		29·7	9
9·34	10·5	11·8	13·4	16·0		18·3	20·5	23·2	25·2	29·6		31·4	10
10·3	11·5	12·9	14·6	17·3		19·7	21·9	24·7	26·8	31·3		33·1	11
11·3	12·6	14·0	15·8	18·5		21·0	23·3	26·2	28·3	32·9		34·8	12
12·3	13·6	15·1	17·0	19·8		22·4	24·7	27·7	29·8	34·5		36·5	13
13·3	14·7	16·2	18·2	21·1		23·7	26·1	29·1	31·3	36·1		38·1	14
14·3	15·7	17·3	19·3	22·3		25·0	27·5	30·6	32·8	37·7		39·7	15
15·3	16·8	18·4	20·5	23·5		26·3	28·8	32·0	34·3	39·3		41·3	16
16·3	17·8	19·5	21·6	24·8		27·6	30·2	33·4	35·7	40·8		42·9	17
17·3	18·9	20·6	22·8	26·0		28·9	31·5	34·8	37·2	42·3		44·4	18
18·3	19·9	21·7	23·9	27·2		30·1	32·9	36·2	38·6	43·8		46·0	19
19·3	21·0	22·8	25·0	28·4		31·4	34·2	37·6	40·0	45·3		47·5	20
20·3	22·0	23·9	26·2	29·6		32·7	35·5	38·9	41·4	46·8		49·0	21
21·3	23·0	24·9	27·3	30·8		33·9	36·8	40·3	42·8	48·3		50·5	22
22·3	24·1	26·0	28·4	32·0		35·2	38·1	41·6	44·2	49·7		52·0	23
23·3	25·1	27·1	29·6	33·2		36·4	39·4	43·0	45·6	51·2		53·5	24
24·3	26·1	28·2	30·7	34·4		37·7	40·6	44·3	46·9	52·6		54·9	25
25·3	27·2	29·2	31·8	35·6		38·9	41·9	45·6	48·3	54·1		56·4	26
26·3	28·2	30·3	32·9	36·7		40·1	43·2	47·0	49·6	55·5		57·9	27
27·3	29·2	31·4	34·0	37·9		41·3	44·5	48·3	51·0	56·9		59·3	28
28·3	30·3	32·5	35·1	39·1		42·6	45·7	49·6	52·3	58·3		60·7	29
29·3	31·3	33·5	36·3	40·3		43·8	47·0	50·9	53·7	59·7		62·2	30
30·3	32·3	34·6	37·4	41·4		45·0	48·2	52·2	55·0	61·1		63·6	31
31·3	33·4	35·7	38·5	42·6		46·2	49·5	53·5	56·3	62·5		65·0	32
32·2	34·4	36·7	39·6	43·7		47·4	50·7	54·8	57·6	63·9		66·4	33
33·3	35·4	37·8	40·7	44·9		48·6	52·0	56·1	59·0	65·2		67·8	34
34·3	36·5	38·9	41·8	46·1		49·8	53·2	57·3	60·3	66·6		69·2	35
35·3	37·5	39·9	42·9	47·2		51·0	54·4	58·6	61·6	68·0		70·6	36
36·3	38·5	41·0	44·0	48·4		52·2	55·7	59·9	62·9	69·3		72·0	37
37·3	39·6	42·0	45·1	49·5		53·4	56·9	61·2	64·2	70·7		73·4	38
38·3	40·6	43·1	46·2	50·7		54·6	58·1	62·4	65·5	72·1		74·7	39
39·3	41·6	44·2	47·3	51·8		55·8	59·3	63·7	66·8	73·4		76·1	40
40·3	42·7	45·2	48·4	52·9		56·9	60·6	65·0	68·1	74·7		77·5	41
41·3	43·7	46·3	49·5	54·1		58·1	61·8	66·2	69·3	76·1		78·8	42
42·3	44·7	47·3	50·5	55·2		59·3	63·0	67·5	70·6	77·4		80·2	43
43·3	45·7	48·4	51·6	56·4		60·5	64·2	68·7	71·9	78·7		81·5	44
44·3	46·8	49·5	52·7	57·5		61·7	65·4	70·0	73·2	80·1		82·9	45
45·3	47·8	50·5	53·8	58·6		62·8	66·6	71·2	74·4	81·4		84·2	46
46·3	48·8	51·6	54·9	59·8		64·0	67·8	72·4	75·7	82·7		85·6	47
47·3	49·8	52·6	56·0	60·9		65·2	69·0	73·7	77·0	84·0		86·9	48
48·3	50·9	53·7	57·1	62·0		66·3	70·2	74·9	78·2	85·4		88·2	49
49·3	51·9	54·7	58·2	63·2		67·5	71·4	76·2	79·5	86·7		89·6	50

α / n	0·9995	0·999	0·995	0·990	0·975	0·95	0·90	0·80	0·70	0·60	0·50
51	24·1	25·4	28·7	30·5	33·2	35·6	38·6	42·4	45·3	47·8	50·3
52	24·8	26·1	29·5	31·2	34·0	36·4	39·4	43·3	46·2	48·8	51·3
53	25·5	26·8	30·2	32·0	34·8	37·3	40·3	44·2	47·2	49·8	52·3
54	26·2	27·5	31·0	32·8	35·6	38·1	41·2	45·1	48·1	50·8	53·3
55	26·9	28·2	31·7	33·6	36·4	39·0	42·1	46·0	49·1	51·7	54·3
56	27·6	28·9	32·5	34·3	37·2	39·8	42·9	47·0	50·0	52·7	55·3
57	28·2	29·6	33·2	35·1	38·0	40·6	43·8	47·9	51·0	53·7	56·3
58	28·9	30·3	34·0	35·9	38·8	41·5	44·7	48·8	51·9	54·7	57·3
59	29·6	31·0	34·8	36·7	39·7	42·3	45·6	49·7	52·9	55·6	58·3
60	30·3	31·7	35·5	37·5	40·5	43·2	46·5	50·6	53·8	56·6	59·3
61	31·0	32·5	36·3	38·3	41·3	44·0	47·3	51·6	54·8	57·6	60·3
62	31·7	33·2	37·1	39·1	42·1	44·9	48·2	52·5	55·7	58·6	61·3
63	32·5	33·9	37·8	39·9	43·0	45·7	49·1	53·5	56·7	59·6	62·3
64	33·2	34·6	38·6	40·6	43·8	46·6	50·0	54·3	57·6	60·5	63·3
65	33·9	35·4	39·4	41·4	44·6	47·4	50·9	55·3	58·6	61·5	64·3
66	34·6	36·1	40·2	42·2	45·4	48·3	51·8	56·2	59·5	62·5	65·3
67	35·3	36·8	40·9	43·0	46·3	49·2	52·7	57·1	60·5	63·5	66·3
68	36·0	37·6	41·7	43·8	47·1	50·0	53·5	58·0	61·4	64·4	67·3
69	36·7	38·3	42·5	44·6	47·9	50·9	54·4	59·0	62·4	65·4	68·3
70	37·5	39·0	43·3	54·4	48·8	51·7	55·3	59·9	63·3	66·4	69·3
71	38·2	39·8	44·1	46·2	49·6	52·6	56·2	60·8	64·3	67·4	70·3
72	38·9	40·5	44·8	47·1	50·4	53·5	57·1	61·8	65·3	68·4	71·3
73	39·6	41·3	45·6	47·9	51·3	54·3	58·0	62·7	66·2	69·3	72·3
74	40·4	42·0	40·4	48·7	52·1	55·2	58·9	63·6	67·2	70·3	73·3
75	41·1	42·8	47·2	49·5	52·9	56·1	59·8	64·5	68·1	71·3	74·3
76	41·8	43·5	48·0	50·3	53·8	56·9	60·7	65·5	69·1	72·3	75·3
77	42·6	44·3	48·8	51·1	54·6	57·8	61·6	66·4	70·0	73·2	76·3
78	43·3	45·0	49·6	51·9	55·5	58·7	62·5	67·3	71·0	74·2	77·3
79	44·1	45·8	50·4	52·7	56·3	59·5	63·4	68·3	72·0	75·2	78·3
80	44·8	46·5	51·2	53·5	57·2	60·4	64·3	69·2	72·9	76·2	79·3
81	45·5	47·3	52·0	54·4	58·0	61·3	65·2	70·1	73·9	77·2	80·3
82	46·3	48·0	52·8	55·2	58·8	62·1	66·1	71·1	74·8	78·1	81·3
83	47·0	48·8	53·6	56·0	59·7	63·0	67·0	72·0	75·8	79·1	82·3
84	47·8	49·6	54·4	56·8	60·5	63·9	67·9	72·9	76·8	80·1	83·3
85	48·5	50·3	55·2	57·6	61·4	64·7	68·8	73·9	77·7	81·1	84·3
86	49·3	51·1	56·0	58·5	62·2	65·6	69·7	74·8	78·7	82·1	85·3
87	50·0	51·9	56·8	59·3	63·1	66·5	70·6	75·7	79·6	83·0	86·3
88	50·8	52·6	57·6	60·1	63·9	67·4	71·5	76·7	80·6	84·0	87·3
89	51·5	53·4	58·4	60·9	64·8	68·2	72·4	77·6	81·6	85·0	88·3
90	52·3	54·2	59·2	61·8	65·6	69·1	73·3	78·6	82·5	86·0	89·3
91	53·0	54·9	60·0	62·6	66·5	70·0	74·2	79·5	83·5	87·0	90·3
92	53·8	55·7	60·8	63·4	67·4	70·9	75·1	80·4	84·4	88·0	91·3
93	54·5	56·5	61·6	64·2	68·2	71·8	76·0	81·4	85·5	88·9	92·3
94	55·3	57·2	62·4	65·1	69·1	72·6	76·9	82·3	86·4	89·9	93·3
95	56·1	58·0	63·2	65·9	69·9	73·5	77·8	83·2	87·3	90·9	94·3
96	56·8	58·8	64·1	66·7	70·8	74·4	78·7	84·2	88·3	91·9	95·3
97	57·6	59·6	64·9	67·6	71·6	75·3	79·6	85·1	89·2	92·9	96·3
98	58·4	60·4	65·7	68·4	72·5	76·2	80·5	86·1	90·2	93·8	97·3
99	59·1	61·1	66·5	69·2	73·4	77·0	81·4	87·0	91·2	94·8	98·3
100	59·9	61·9	67·3	70·1	74·2	77·9	82·4	87·9	92·1	95·8	99·3

0·40	0·30	0·20	0·10	0·05	0·025	0·01	0·005	0·001	0·0005	α / n
52·9	55·8	59·2	64·3	68·7	72·6	77·4	80·7	88·0	90·9	51
53·9	56·8	60·3	65·4	69·8	73·8	78·6	82·0	89·3	92·2	52
55·0	57·9	61·4	66·5	71·0	75·0	79·8	83·3	90·6	93·5	53
56·0	58·9	62·5	67·7	72·2	76·2	81·1	84·5	91·9	94·8	54
57·0	60·0	63·6	68·8	73·3	77·4	82·3	85·7	93·2	96·2	55
58·0	61·0	64·7	69·9	74·5	78·6	83·5	87·0	94·5	97·5	56
59·1	62·1	65·7	71·0	75·6	79·8	84·7	88·2	95·8	98·8	57
60·1	63·1	66·8	72·2	76·8	80·9	86·0	89·5	97·0	100·1	58
61·1	64·2	67·9	73·3	77·9	82·1	87·2	90·7	98·3	101·4	59
62·1	65·2	69·0	74·4	79·1	83·3	88·4	92·0	99·6	102·7	60
63·2	66·3	70·0	75·5	80·2	84·5	89·6	93·2	100·9	104·0	61
64·2	67·3	71·1	76·6	81·4	85·7	90·8	94·4	102·2	105·3	62
65·2	68·4	72·2	77·7	82·5	86·8	92·0	95·6	103·4	106·6	63
66·2	69·4	73·3	78·9	83·7	88·0	93·2	96·9	104·7	107·9	64
67·2	70·5	74·4	80·0	84·8	89·2	94·4	98·1	106·0	109·2	65
68·3	71·5	75·4	81·1	86·0	90·3	95·6	99·3	107·3	110·5	66
69·3	72·6	76·5	82·2	87·1	91·5	96·8	100·6	108·5	111·7	67
70·3	73·6	77·6	83·3	88·3	92·7	98·0	101·8	109·8	113·0	68
71·3	74·6	78·6	84·4	89·4	93·9	99·2	103·0	111·1	114·3	69
72·4	75·7	79·7	85·5	90·5	95·0	100·4	104·2	112·3	115·6	70
73·4	76·7	80·8	86·6	91·7	96·2	101·6	105·4	113·6	116·9	71
74·4	77·8	81·9	87·7	92·8	97·4	102·8	106·6	114·8	118·1	72
75·4	78·8	82·9	88·8	93·9	98·5	104·0	107·9	116·1	119·4	73
76·4	79·9	84·0	90·0	95·1	99·7	105·2	109·1	117·3	120·7	74
77·5	80·9	85·1	91·1	96·2	100·8	106·4	110·3	118·6	121·9	75
78·5	82·0	86·1	92·2	97·4	102·0	107·6	111·5	119·9	123·2	76
79·5	83·0	87·2	93·3	98·5	103·2	108·8	112·7	121·1	124·5	77
80·5	84·0	88·3	94·4	99·6	104·3	110·0	113·9	122·3	125·7	78
81·5	85·1	89·3	95·5	100·7	105·5	111·1	115·1	123·6	127·0	79
82·6	86·1	90·4	96·6	101·9	106·6	112·3	116·3	124·3	128·3	80
83·6	87·2	91·5	97·7	103·0	107·8	113·5	117·5	126·1	129·5	81
84·6	88·2	92·5	98·8	104·1	108·9	114·7	118·7	127·3	130·8	82
85·6	89·2	93·6	99·9	105·3	110·1	115·9	119·9	128·6	132·0	83
86·6	90·3	94·7	101·0	106·4	111·2	117·1	121·1	129·8	133·3	84
87·7	91·3	95·7	102·1	107·5	112·4	118·2	122·3	131·0	134·5	85
88·7	92·4	96·8	103·2	108·6	113·5	119·4	123·5	132·3	135·8	86
89·7	93·4	97·9	104·3	109·8	114·7	120·6	124·7	133·5	137·0	87
90·7	94·4	98·9	105·4	110·9	115·8	121·8	125·9	134·7	138·3	88
91·7	95·5	100·0	106·5	112·0	117·0	122·9	127·1	136·0	139·5	89
92·8	96·5	101·1	107·6	113·1	118·1	124·1	128·3	137·2	140·8	90
93·8	97·6	102·1	108·7	114·3	119·3	125·3	129·5	138·4	142·0	91
94·8	98·6	103·2	109·8	115·4	120·4	126·5	130·7	139·7	143·3	92
95·8	99·6	104·2	110·9	116·5	121·6	127·6	131·9	140·9	144·5	93
96·8	100·7	105·3	111·9	117·6	122·7	128·8	133·1	142·1	145·8	94
97·9	101·7	106·4	113·0	118·8	123·9	130·0	134·2	143·3	147·0	95
98·9	102·8	107·4	114·1	119·9	125·0	131·1	135·4	144·6	148·2	96
99·9	103·8	108·5	115·2	121·0	126·1	132·3	136·6	145·8	149·5	97
100·9	104·8	109·5	116·3	122·1	127·3	133·5	137·8	147·0	150·7	98
101·9	105·9	110·6	117·4	123·2	128·4	134·6	139·0	148·2	151·9	99
102·9	106·9	111·7	118·5	124·3	129·6	135·8	140·2	149·4	153·2	100

Appendix 3 Microelectronics failure rates

The following table gives rates per million hours showing the highest and lowest values likely to be quoted in data bases. The middle column is the geometric mean (Section 4.3). Each group of three columns is labelled for a junction temperature range in degrees Centigrade. The following multipliers apply:

	MULTIPLIER

QUALITY

Normal commercial procurement	2
Procured to some agreed specification and	1
Quality management system	
100% screening and burn-in	0.4

ENVIRONMENT

Dormant (little stress)	0.1
Benign (e.g. air-conditioned)	0.5
Fixed ground (no adverse vibration, temperature cycling etc.)	1
Mobile/portable	4

PACKAGING

Ceramic	1
Plastic	1 for quality factor 0.4
	2 for quality factors 1 or 2

Logic	<40			40–62		
Bipolar SRAM 64k bits	0.03	0.06	0.13	0.05	0.08	0.13
Bipolar SRAM 256k bits	0.04	0.14	0.50	0.09	0.21	0.50
Bipolar PROM/ROM 256k bits	0.02	0.02	0.03	0.03	0.03	0.03
Bipolar PROM/ROM 16k bits	0.03	0.03	0.04	0.03	0.04	0.06
MOS SRAM 16k bits	0.02	0.02	0.03	0.02	0.03	0.05
MOS SRAM 4m bits	0.08	0.19	0.44	0.20	0.30	0.44
MOS DRAM 64k bits	0.02	0.02	0.02	0.02	0.02	0.03
MOS DRAM 16m bits	0.05	0.11	0.23	0.09	0.14	0.23
MOS EPROM 16k bits	0.03	0.05	0.07	0.04	0.05	0.07
MOS EPROM 8m bits	0.06	0.13	0.30	0.07	0.14	0.30

	62–87			>87		
Bipolar SRAM 64k bits	0.13	0.14	0.15	0.13	0.25	0.48
Bipolar SRAM 256k bits	0.30	0.39	0.50	0.50	0.70	0.96
Bipolar PROM/ROM 256k bits	0.03	0.05	0.08	0.03	0.03	0.03
Bipolar PROM/ROM 16k bits	0.03	0.07	0.15	0.03	0.12	0.47
MOS SRAM 16k bits	0.02	0.05	0.13	0.02	0.09	0.38
MOS SRAM 4m bits	0.44	0.59	0.80	0.44	1.09	2.70
MOS DRAM 64k bits	0.02	0.03	0.05	0.02	0.05	0.13
MOS DRAM 16m bits	0.23	0.25	0.28	0.23	0.46	0.92
MOS EPROM 16k bits	0.04	0.05	0.07	0.04	0.05	0.07
MOS EPROM 8m bits	0.14	0.20	0.30	0.30	0.33	0.36

	<40			40–62		
Linear Bipolar 50 tr	0.01	0.01	0.02	0.01	0.02	0.03
Linear MOS 50 tr	0.02	0.03	0.04	0.03	0.03	0.04
Logic Bipolar 50 gate	0.01	0.01	0.02	0.01	0.01	0.02
Logic Bipolar 500 gate	0.01	0.01	0.02	0.01	0.02	0.03
Logic MOS 50 gate	0.01	0.01	0.02	0.01	0.02	0.03
Logic MOS 500 gate	0.01	0.02	0.03	0.01	0.02	0.05
MicroProc Bipolar 8 bit	0.01	0.03	0.07	0.01	0.04	0.14
MicroProc Bipolar 16 bit	0.01	0.03	0.08	0.01	0.05	0.23
MicroProc Bipolar 32 bit	0.01	0.03	0.11	0.01	0.06	0.40
MicroProc MOS 8 bit	0.02	0.04	0.10	0.02	0.05	0.14
MicroProc MOS 16 bit	0.02	0.06	0.18	0.02	0.08	0.30
MicroProc MOS 32 bit	0.02	0.08	0.32	0.02	0.10	0.55
ASIC/PLA/FPGA Bip'lr 1k gate	0.05	0.06	0.07	0.05	0.07	0.12
ASIC/PLA/FPGA MOS 1k gate	0.05	0.05	0.06	0.05	0.05	0.06
GaAs/MMIC 100 element	0.06	0.06	0.07	0.06	0.06	0.07

	62–87			>87		
Linear Bipolar 50 tr	0.01	0.03	0.10	0.01	0.06	0.34
Linear MOS 50 tr	0.05	0.07	0.10	0.03	0.10	0.34
Logic Bipolar 50 gate	0.01	0.02	0.04	0.01	0.03	0.10
Logic Bipolar 500 gate	0.01	0.02	0.06	0.01	0.04	0.18
Logic MOS 50 gate	0.02	0.03	0.04	0.02	0.03	0.06
Logic MOS 500 gate	0.02	0.03	0.06	0.02	0.04	0.10
MicroProc Bipolar 8 bit	0.01	0.07	0.54	0.01	0.14	2.00
MicroProc Bipolar 16 bit	0.01	0.10	1.00	0.01	0.20	4.00
MicroProc Bipolar 32 bit	0.01	0.14	2.00	0.01	0.28	7.70
MicroProc MOS 8 bit	0.02	0.07	0.26	0.02	0.10	0.50
MicroProc MOS 16 bit	0.02	0.10	0.50	0.02	0.14	1.00
MicroProc MOS 32 bit	0.02	0.14	1.00	0.02	0.20	2.00
ASIC/PLA/FPGA Bip'lr 1k gate	0.05	0.14	0.40	0.05	0.26	1.40
ASIC/PLA/FPGA MOS 1k gate	0.05	0.05	0.06	0.05	0.06	0.07
GaAs/MMIC 100 element	0.06	0.06	0.07	0.06	0.06	0.07

Appendix 4 General failure rates

This appendix, which is an extract from an early version of FARADIP.THREE, provides some failure rates. The multiplying factors for quality and environment, together with an explanation of the columns, are given in Appendix 3.

Item	*Failure rate in failures per million hours*		
Accelerometer	10		30
Air Compressor	70		250
Air Supply (Instrument)	5	6	10
Alarm Bell	2		10
Alarm Circuit (Simple)	4		
(Panel)	45		
Alarm Siren	1	6	20
Alternator	1		9
Analyser – CO_2	100		500
– Conductivity	500	1500	2000
– Dewpoint	100		200
– Geiger	15		
– Hydrogen	400		100
– Oxygen	50	60	200
– pH	650		
– Scintillation	20		
– Bourdon/Geiger	5		
– H_2S	100		200
Antenna	1		5
Attenuator	0.01		
Battery – Lead-Acid	0.5	1	3
– Ni-Cd/Ag-Zn	0.2	1	3
– Lead-Acid (vehicle) per million miles	30		
– Dry primary	1		30
Battery charger			
– Simple rectifier	2		
– Stabilized/float	10		
– Motor generator	100		
Battery Lead	3		

Item			
Bearings – Ball, light	0.1	1	10
– Ball, heavy	2		20
– Roller	0.3		5
– Sleeve	0.5		5
– Jewel	0.4		
– Brush	0.5		
– Bush	0.05		0.4
Bellows, simple expandable	2	5	10
Belts	4		50
Busbars 11 kV	0.02		**0.2**
– 3.3 kV	0.05		2
– 415 V	0.6		2
Cable (Power) per km			
– Overhead < 600 V	0.5		
600-15 kV	5		15
> 33 kV	3		7
– Underground < 600 V	2		
600-15 kV	2		
– Subsea	2.5		
Capacitors			
– Paper	0.001	**0.15**	
– Plastic	0.001	**0.01**	0.05
– Mica	0.002	0.03	**0.1**
– Glass	0.002		
– Ceramic	0.0005	0.1	
– Tant. sol.	0.005	0.1	
– Tant. non-sol.	0.001	**0.01**	0.1
– Alumin. (gen.)	0.3		
– Variable	0.005	**0.1**	2
Card Reader	**150**		4000
Circuit Breaker			
– < 600 V or A	0.5		1.5
– > 3 kV	0.5		2
– > 100 kV	3		10
Clutch – Friction	0.5		3
– Magnetic	2.5		**6**
Compressor			
– Centrifugal, turbine driven	150		
– Reciprocating turbine driven	500		
– Electric motor driven	100		300
Computer			
– Mainframe	4000		8000
– Mini	100	200	500
– Micro (CPU)	30		100
– PLC	20		50
Connections			
– Hand solder	0.0002		0.003
– Flow solder	0.0003		0.001
– Weld	0.002		

– Wrapped	0.00003		0.001	
– Crimped	0.0003		0.007	
– Power cable	0.05		0.4	
– Plate th. hl.	0.0003			
Connectors				
– Coaxial	0.02		0.2	
– PCB	0.0003		0.1	
– pin	0.001		0.1	
– r.f.	0.05			
– pneumatic	1			
– DIL	0.001			
Counter (mech.)	0.2	2		
Crystal, Quartz	0.02	0.1	0.2	
Detectors				
– Gas, pellistor	3		8	
– Smoke, ionization	2		6	
– Ultra-violet	5		15	
– Rate of rise (temp.)	3		9	
– Temperature level	0.2	2	8	
– Fire, Wire/rod	10			
Diesel Engine	**300**	6000		
Diesel Generator	125		4000 (0.97 start)	
Diodes – Si, high power	**0.1**	0.2		
– Si, low power	0.01	**0.04**	0.1	
– Zener	0.005	0.03	**0.1**	
– Varactor	0.06		0.3	
– SCR (Thyristor)	0.01		0.5	
Disk Memory	100	500	2000	
Electricity Supply	100			
Electropneumatic Converter (I/P)	2		4	
Fan	2		50	
Fibre Optics				
– Connector	0.1			
– Cable/km	0.1			
– LED	0.2		0.5	
– Laser	0.3		0.5	
– Si Avalanche photodiode	0.2			
– Pin Avalanche photodiode	0.02			
– Optocoupler	0.02		0.1	
Filter (Blocked)	0.5	1	10	
(Leak)	0.5	1	10	
Fire Sprinkler (spurious)	0.05	**0.1**	0.5	0.02 probability of non-operation
Fire Water Pump System	150	200	800	
Flow Instruments				
– Transmitter	1	**5**	20	
– Controller	25		50	
– DP sensor	80		200	
– Switch	4		40	

	– Rotary meter	5	15		
Fuse		0.02		0.5	(Mobile 2-20)
Gaskets		0.05	**0.4**	3	
Gear	– per mesh	0.05	**0.5**	1	
	– Assembly	10		50	Proportional to size
Generator					
	– a.c.	3		30	
	– d.c.	1		10	
	– Turbine set	10	**200**	800	
	– Motor set	30		70	
	– Diesel set	125		4000	(Standby 8-200)
Hydraulic Equipment					
	– Accumulator/damper	**20**	200		
	– Actuator	15			
	– Piston	1			
	– Motor	5			
Inductor (l.f., r.f.)		0.2		0.5	
Joints	– Pipe	0.5			
	– O ring	0.2		0.5	
Lamps	– Filament	0.05	**1**	10	
	– Neon	0.1	0.2	1	
LCD (per character)		0.05			
(per device)		2.5			
LED	– Indicator	0.06		0.3	
	– Numeral (per char.)	0.01		0.1	
Level Instruments					
	– Switch	2	**5**	20	
	– Controller	4		20	
	– Transmitter	10		20	
	– Indicator	1		10	
Lines (Communications)					
	– Speech channel, land	100		250	
	– Coaxial/km	1.5			
	– Subsea/km	2.4			
Load Cell		100		400	
Loudspeaker		10			
Magnetic Tape Unit, incl. drive		200		500	
Meter (Moving Coil)		1		5	
Microwave Equipment					
	– Fixed element	0.01			
	– Tuned element	0.1			
	– Detector/mixer	0.2			
	– Waveguide, fixed	1			
	– Waveguide, flexible	2.5			
Motor (Electrical)					
	– a.c.	1	**5**	20	
	– d.c.	5	**15**		
	– Starter	4		10	

Optodevices				See Fibre Optics
Photoelectric Cell	15			
Pneumatic Equipment				
– Connector	1.5			
– Controller	1		2	Open or short
– Controller	10		20	Degraded
– I/P converter	2		10	
– Pressure relay	20			
Power Supply				
– d.c./d.c. converter	2	5	20	
– a.c./d.c. stabilized	5	20	100	If possible carry out FMEA
Pressure Instruments				
– Switch	1	**5**	40	
– Sensor	2		10	
– Indicator	1	**5**	10	
– Controller	1	**10**	30	1 catastrophic, 20 degraded
– Transmitter (P/I) (I/P)	5		20	
Printed Circuit Boards				
– Single sided	0.02			
– Double (plated through)	0.01		0.3	
– Multilayer	0.07		0.1	
Printer (Line)	300		1000	
Pumps – Centrifugal	10	50	100	
– Boiler	100		700	
– Fire water – diesel	200		3000	
– electr.	200		500	
– Fuel	3		180	
– Oil lubrication	6		70	
– Vacuum	10		25	
Pushbutton	0.1	0.5	10	
Rectifier (Power)	3		5	
Relays – Armature general	0.2		0.4	
– Crystal can	0.15			
– Heavy duty	2		5	
– Polarized	0.8			
– Reed	0.002	**0.2**	2	
– BT	0.02		0.07	
– Contactor	1		6	
– Power	1		16	
– Thermal	0.5		10	
– Time delay	0.5	**2**	10	
– Latching	0.02		1.5	
Resistors – Carbon comp	0.001		0.006	
– Carbon film	0.001		0.05	
– Metal oxide	0.001	**0.004**	0.05	
– Wire wound	0.001	**0.005**	0.5	
– Networks	0.05		0.1	

– Variable WW	0.02	**0.05**	0.5	
– Variable comp.	0.5		1.5	
Solenoid	0.4	**1**	4	
Stepper Motor	0.5		5	
Surge Arresters				
– > 100 kV	0.5		1.5	
– low power	0.003		0.02	
Switches (per contact)				
– Micro	0.1		1	
– Toggle	0.03		1	
– DIL	0.03	0.5	1.8	
– Key (low power)	0.003		2	
(high power)	5		10	
– Pushbutton	**0.2**	1	10	
– Rotary	0.05		0.5	
– Thermal Delay	0.5		3	
Synchros and Resolvers	**3**		15	
Temperature Instruments				
– Sensor	0.2		10	
– Switch	**3**		20	
– Pyrometer	250		1000	
– Transmitter	10			
– Controller	20		40	
Thermionic Tubes				
– Diode	5	20	70	
– Triode and Pentode	20	30	100	
– Thyratron	50			
Thermocouple/Thermostat	1	**10**	20	
Timer (electromech.)	2	15	40	
Transformers				
– Signal	0.005	0.2	**0.3**	
– Mains	0.03	0.4	**3**	
– ≥ 415 V	**0.4**	1	7	
Transistors				
– Si npn low power	0.01	**0.05**	0.2	
– Si npn high power	0.1		0.4	
– Si FET low power	0.05			
– Si FET high power	0.1			
Turbine, Steam	30	40		
TV Receiver	2.3			1984 figure
Valves (Mechanical, Hydraulic, Pneumatic, Gas (not high temp. nor corrosive substances))				
– Ball	0.2	3	10	
– Butterfly	1	**20**	30	
– Diaphragm (single)	2.6	**10**	20	
– Gate	1	10	30	
– Needle	1.5	20		
– Non-return	1		20	

– Plug	1		18	
– Relief	2		8	
– Globe	0.2		2	
– Solenoid	1		8	De-energize to trip
– Solenoid	8	20		Energize to trip
Valve diaphragm	1	5		
VDU	10	200	500	

Appendix 5 Failure mode percentages

Just as the failure rates in the preceding tables must vary according to a large number of parameters, then so must the relative percentages of the different failure modes. However, the following figures will provide the reader with some general information which may be of assistance in carrying out a Failure Mode Analysis where no more specific data are available. The total item failure rate may be multiplied by the appropriate failure mode percentage in order to estimate the mode failure rate.

Item	Mode	Percentage
Battery	Catastrophic Open	10
	Catastrophic Short	20
	Leak	20
	Low Output	50
Bearing	Binding	40
	Worn	60
Capacitor		
Electrolytic	Open Circuit	20
	Short Circuit	80
Mica, Ceramic, Glass, Paper	Open Circuit	1
	Short Circuit	99
Plastic	Open Circuit	50
	Short Circuit	50
Circuit Breaker	Arcing and Damage	10
	Fail to Close	5
	Fail to Open	40
	Spurious Open	45
Clutch (Mechanical)	Bind	55
	Slip	45
Connection (Solder)	Break	50
	Dry	40
	No Solder	10
Connector	High Resistance	10
	Intermittent	20
	Open Circuit	60
	Short	10
Diesel Engine	Air and Fuel	23
	Blocks and Heads	7
	Elec., Start, Battery	1

	Lube and Cooling		23
	Misc. and Seals		16
	Moving Mech. Parts		30
Diode (Junction)	High Reverse		60
	Open		25
	Short		15
Diode (Zener)	Open		50
	Short		50
Fuse	Fails to Open		15
	Opens		10
	Slow to Open		75
Gear	Binding		80
	No Transmission		20
Generator	Drift or Intermittent		80
	Loss of Output		20
Inductor	Open		75
	Short		25
Lamp	Open		100
Meter (Moving Coil)	Drift		30
	No Reading		70
Microelectronics			
Digital	High		40
	Loss of Logic		20
	Low		40
Linear	Drift		10
	High or Low		10
	No Output		80
Motor	Failed – Brush	15	
	– Commutator	10	
	– Lube	15	65
	– Rotor	10	
	– Stator	15	
	Performance (Degraded)		
	– Brush	15	
	– Commutator	5	35
	– Lube	15	
Pump	Leak		50
	No Transmission		50
Relay	Coil		10
	Contact		90
Relay, Contact	Fail to Operate		90
	Fail to Release		10
Resistor (Comp.)	Open		50
	Drift		50
(Film)	Open		50
	Drift		50
(Var.)	Open		40
	Intermittent		60
(Wire)	Open		90

	Short		10
SCR	Open		2
	Short		98
Switch (Micro)	High Resistance		60
	No Function		10
	Open		30
(Pushbutton)	Open		80
	Short		20
Transformer	Open – Primary	50	60
	– Secondary	10	
	Short – Primary	30	40
	– Secondary	10	
Transistor	High Leakage		20
	Low Gain		20
	Open Circuit		30
	Short Circuit		30
Valve (Mechanical)	Blocking		5
	External Leak		15
	Passing (Internal)		60
	Sticking		20
Valve Actuator	Fail		10
	Spurious		90

Note: Can be Spurious Open or Spurious Close, Fail Open or Fail Close, depending on the hydraulic logic

Appendix 6 Human error rates

In more general Reliability work, risk analysis is sometimes involved. Also, system MTBF calculations often take account of the possibilities of human error. A number of studies have been carried out in the UK and the USA which attempt to quantify human error rates. The following is an overview of the range of error rates which apply.

It must be emphasized that these are broad guidelines. In any particular situation the human-response reliability will be governed by a number of shaping factors which include:

Environmental factors	– Physical
	– Organizational
	– Personal
Intrinsic error	– Selection of Individuals
	– Training
	– Experience
Stress factors	– Personal
	– Circumstantial

	Error rate (per task)		
	Read/ reason	*Physical operation*	*Everyday yardstick*
Simplest possible task			
Fail to respond to annunciator	0.0001		
Overfill bath			0.00001
Fail to isolate supply (electrical work)		0.0001	
Read single alphanumeric wrongly	0.0002		
Read 5-letter word with good resolution wrongly	0.0003		
Select wrong switch (with mimic diagram)		0.0005	
Fail to notice major cross-roads			0.0005

	Error rate (per task)		
	Read/ reason	*Physical operation*	*Everyday yardstick*
Routine simple task			
Read a checklist or digital display wrongly	0.001		
Set switch (multiposition) wrongly		0.001	
Calibrate dial by potentiometer wrongly		0.002	
Check for wrong indicator in an array	0.003		
Wrongly carry out visual inspection for a defined criterion (e.g. leak)	0.003		
Fail to correctly replace PCB		0.004	
Select wrong switch among similar		0.005	
Read analogue indicator wrongly	0.005		
Read 10-digit number wrongly	0.006		
Leave light on			0.003
Routine task with care needed			
Mate a connector wrongly		0.01	
Fail to reset valve after some related task		0.01	
Record information or read graph wrongly	0.01		
Let milk boil over			0.01
Type or punch character wrongly		0.01	
Do simple arithmetic wrongly	0.01–0.03		
Wrong selection – vending machine			0.02
Wrongly replace a detailed part		0.02	
Do simple algebra wrongly	0.02		
Read 5-letter word with poor resolution wrongly	0.03		
Put 10 digits into calculator wrongly	0.05		
Dial 10 digits wrongly	0.06		
Complicated non-routine task			
Fail to notice adverse indicator when reaching for wrong switch or item	0.1		
Fail to recognize incorrect status in roving inspection	0.1		
New workshift – fail to check hardware, unless specified	0.1		
General (high stress)	0.25		
Fail to notice wrong position of valves	0.5		
Fail to act correctly after 1 min in emergency situation	0.9		

In failure rate terms the incident rate in a plant is likely to be in the range of 20×10^{-6} per h (general human error) to 1×10^{-6} per h (safety-related incident).

Appendix 7 Fatality rates

The following are approximate fatality rates for the UK (summarized from numerous sources) for a number of Occupational, Voluntary, Involuntary and Travel risks. They are expressed as rates, which for small values may be taken as probabilities, on the basis of annual and exposed hours. A rate per year expresses the probability of an individual becoming a fatality in one year, given a normal exposure to the risk in question. A FAFR (Fatal Accident Frequency Rate) is expressed on the basis of the number of expected fatalities per 100 million exposed hours.

	Per year	FAFR	Other
Travel			
Air (Scheduled)	2×10^{-6}		1×10^{-7} per landing
			or 5×10^{-5} per lifetime
			or 2×10^{-10} per km
Train	3×10^{-6}	3–5	1×10^{-9} per km
Bus	1×10^{-4}	4	5×10^{-10} per km
Car	0.5×10^{-4}	50–60	*c*. 3500 per year
			4×10^{-10} per km
Canoe		650	
Gliding		1000	
Motor cycle	2×10^{-2}	500–1050	10^{-7} per km
Water (General)	2×10^{-6}		9×10^{-9} per km
Occupation			
British Industry		2–4 (USA 7)	*c*. 800 per year (UK)
Chemical Industry	5×10^{-5}	4	
Construction	1×10^{-4}		
Construction Erectors		10–70	
Mining (Coal)	1×10^{-4}	10 (USA 30)	
Nuclear	4×10^{-5}		
Railway Shunting	2×10^{-4}	45	
Boxing		20000	
Steeplejack		300	
Boilers (100% exposure)	3×10^{-5}	0.3	
Agriculture	7×10^{-5}	10 (USA 3)	
Mechanical, Manufacturing		8	
Oil and gas extraction	1×10^{-3}		
Furniture		3	

	Per year	FAFR	Other
Clothing/Textiles	2×10^{-5}	0.2	
Electrical Engineering	1×10^{-5}		
Shipping	9×10^{-4}	8	c. 250 per year

Voluntary

	Per year	FAFR	Other
Smoking (20 per day)	500×10^{-5}		
Drinking (3 pints per day)	8×10^{-5}		
Football	4×10^{-5}		
Car Racing	120×10^{-5}		
Rock Climbing	14×10^{-5}	4000	4×10^{-5} per hour
The Pill	2×10^{-5}		
Swimming		1300	

Involuntary

	Per year	FAFR	Other
Earthquake, UK	2×10^{-8}		
Earthquake, California	2×10^{-6}		
Lightning (in UK)	1×10^{-7}		
Skylab strike	5×10^{-12}		
Pressure vessels	5×10^{-8}		
Nuclear (1 km)	1×10^{-7}		
Run over	6×10^{-5}		
Falling aircraft	2×10^{-8}		
Venomous bite	2×10^{-7}		
Petrol/chemical transport	2×10^{-8}		1 in 670 million miles
Leukaemia	8×10^{-5}		
Influenza	2×10^{-4}		
Meteorite	6×10^{-11}		
Firearms/explosive	1×10^{-6}		
Homicide	1×10^{-5}		
Drowning	1×10^{-5}		
Fire	2×10^{-5}		
Poison	1.5×10^{-5}		
Suicide	8×10^{-5}		
Falls	1×10^{-4}		
Staying at Home		1–4	
All accidents	4×10^{-4}		
Electrocution	1.2×10^{-6}		
Cancer	25×10^{-4}		
All accidents	3×10^{-4}		
Natural disasters (general)	2×10^{-6}		
All causes*	1×10^{-2}		

*See *A Healthier Mortality* ISBN 0-952 5072-1-8.

Appendix 8 Answers to exercises

Chapter 2

	a)	b)
1.	114	1.1
2.	0.99	0.42 (0.12*)
3.	10^{-5}	10^{-3}
4.	2.2 10^{-3}	0.18 (0.22*)
5.	Negligible	Negligible
6.	Unavailability $\times$ 2	Unavailability $\times$ 2

* Beware the approximation. λt is large.

Chapter 5

1. Accumulated time $T = 50 \times 100 = 5000$ h. Since the test was time truncated $n = 2(k + 1)$. Therefore,

   (a) $n = 6$, $T = 5000$, $\alpha = 0.4$. From Appendix 2, $\chi^2 = 6.21$

   $$\text{MTBF}_{60\%} = \frac{2T}{\chi^2} = \frac{10\ 000}{6.21} = 1610 \text{ h}$$

   (b) $n = 2$, $T = 5000$, $\alpha = 0.4$. From Appendix 2, $\chi^2 = 1.83$

   $$\text{MTBF}_{60\%} = \frac{2T}{\chi^2} = \frac{10\ 000}{1.83} = 5464 \text{ h}$$

2. If $k = 0$ then $n = 2$ and since confidence level = 90% $\alpha = 0.1$

   Therefore $\chi^2 = 4.61$

   $$\text{MTBF}_{90\%} = 5000 = \frac{2T}{\chi^2} = \frac{2T}{4.61}$$

   Therefore $T = \dfrac{5000 \times 4.61}{2} = 11\ 525 \text{ h}$

   Since there are 50 devices the duration of the test is $\dfrac{11\ 525}{50} = 231$ h.

3. From Figure 5.7. If $c = 0$ and $P_{0-c} = 0.85$ ($\alpha = 0.15$) then $m = 0.17$

   Therefore $T = m\theta = 0.17 \times 1000 = 170$ h

   If MTBF is 500 h then $m = T/\theta = 170/500 = 0.34$ which shows $\beta = 70\%$

   If $c = 5$ then $m = 3.6$ at $P_{0-c} = 0.85$

   Therefore $T = m\theta = 3.6 \times 1000 = 3600$ h

   If MTBF is 500 h then $m = T/\theta = 3600/500 = 7.2$ which shows $\beta = 28\%$

   NB: Do not confuse α meaning $(1 - $ confidence level) with α as producer's risk.

Chapter 6

1. From the example $R(t) = \left[\left(\dfrac{-t}{1110}\right)^{1.5}\right]$

   If $R(t) = 0.95$ Then $\left(\dfrac{t}{1110}\right)^{1.5} = 0.051$

   Therefore $1.5 \log (t/1110) = \log 0.051$

   Therefore $\log (t/1110) = -1.984$

   Therefore $t/1110 = 0.138$

   Therefore $t = 153$ h

2. Using the table of median ranks, sample size 10, as given in Chapter 6, plot the data and verify that a straight line is obtained.

   Note that $\beta = 2$ and that $\eta = 13\ 000$ h

   Therefore

$$R(t) = \exp\left[\left(\dfrac{-t}{13\ 000}\right)^{2}\right]$$

   and

   MTBF $= 0.886 \times 13\ 000 = 11\ 500$ h

Chapter 7

1. $R(t) = e^{-\lambda t} [e^{-\lambda t} - e^{-\lambda t}]$

    $= 2e^{-\lambda t} - 3^{-3\lambda t}$

   $MTBF = \displaystyle\int_{0}^{\infty} R(t)\ dt = \dfrac{1}{\lambda} - \dfrac{1}{3\lambda} = \dfrac{2}{3\lambda}$

   NB: Not a constant failure rate system despite λ being constant.

2. This is a conditional redundancy problem. Consider the reliability of the system if (a) B does fail and (b) B does not fail. The following two block diagrams describe the equivalent systems for these two possibilities.

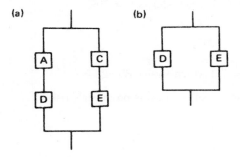

Using Bayes theorem the reliability is given as:

Reliability of diagram (a) $\times$ probability that B fails (i.e. $1 - R_b$)

    PLUS

Reliability of diagram (b) $\times$ probability that B does not fail (i.e. R_b)

Therefore System Reliability

$$= [R_a R_d + R_c R_e - R_a R_d R_c R_e]\,(1 - R_b) + [R_d + R_e - R_d R_e]\,R_b$$

Chapter 9

1(a) *Loss of supply* – Both streams have to fail, i.e. the streams are in parallel, hence the reliability block diagram is

$$R = 1 - (1 - R_s)\,(1 - R_s)$$

where R_s is the reliability of each stream from Section 7.3

$$R = 1 - (1 - 0.885)\,(1 - 0.885) = \underline{0.9868}$$

1(b) *Overpressure* – occurs if *either* stream fails open, hence the streams are in series from a reliability point of view, and the block diagram is:

$$R = R_s{}^2 \text{ where } R_s \text{ is the reliability of each stream from Section 7.4.2}$$

$$R = (0.999)^2 = \underline{0.998}$$

Notes

The twin stream will *reduce* the risk of loss of supply, but *increase* the risk of over-pressure.

The same principles can be used to address more realistic complex systems with non-return valves, slam shuts pressure transducers, etc.

R will be increased if loss of supply in one stream can be detected and repaired while the other stream supplies. The down time of a failed stream is then relevant to the calculation and different.

2.

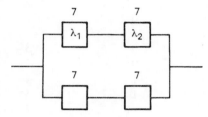

λ (Stream) $= \lambda_1 + \lambda_2 = 14 \ 10^{-6}$ ph

Thus:

Failure Rate $\approx 2\lambda^2 MDT$ where MDT $= \frac{1}{2}$ of 2 weeks

$$= 2(14 \ 10^{-6})^2 \times 168$$

$$= 0.0659 \ 10^{-6}$$

MTBF $= 1/\lambda = 1733$ years

3.

The overall Unavailability is 0.01. Calculating the Unavailability for each cutset:

MOTOR 0.0084   (84%)

PUMP 0.00144   (14%)

PSU and STANDBY 0.0002   (2%)

UV DETECTOR and PANEL (negl)

Note that the ranking, and the percentage contributions, are not the same as for failure rate.

Chapter 12

1.

Cumulative hours	Failures	Anticipated	Deviation	Cusum
3000	1	1	0	0
6000	2	1	1	1
9000	2	1	1	2
12 000	1	1	0	2
15 000	2	1	1	3
18 000	0	1	−1	2
21 000	1	1	0	2
24 000	2	1	1	3
27 000	1	1	0	3
30 000	1	1	0	3
33 000	2	1	1	4
36 000	2	1	1	5
39 000	0	1	−1	4
42 000	1	1	0	4
45 000	0	1	−1	3
48 000	0	1	−1	2
51 000	0	1	−1	1
54 000	0	1	−1	0
57 000	0	1	−1	−1
60 000	0	1	−1	−2

2.

$T_1 = 50 \times 8760 \times 0.25 = 109\,500$

$\theta_1 = 109\,500/20 = 5475$ hrs

$T_2 = 109\,500 + 100 \times 8760 \times 0.25 = 328\,500$

$\theta_2 = 328\,500/35 = 9386$ hrs

$\theta_2/\theta_1 = (T_2/T_1)^\alpha$ Therefore $1.714 = 3^\alpha$ Therefore $\alpha = 0.5$

$\theta = k\,T^\alpha$ So $5475 = k \times 331$ Therefore $k = 16.5$

For MTBF to be 12 000, $T^{0.5} = 12\,000/16.5$ so $T = 528\,900$ hours.

which is another 200 400 hours.

which will take c 2000 hours with the number on trial.

If $\alpha = 0.6$, k changes as follows:

$k\,(328\,500)^{0.6} = 9386$ Therefore $k = 4.6$

Now MTBF is 12 000 at $T^{0.6} = 12\,000/4.6$ so $T = 491\,800$ hours.

which is another 163 300 hours.

which will take c 1600 hours with the number on trial.

Appendix 9 Bibliography

BOOKS

Carter, A. D. S., *Mechanical Reliability*, 2nd edn, Macmillan, London (1986)

Collins, J. A., *Failure of Materials in Mechanical Design*, Wiley, New York (1981)

Fullwood, R. F., *Probabilistic Safety Assessment in the Chemical and Nuclear Industries*, Butterworth-Heinemann, Oxford (1999) ISBN 0 7506 7208 0

Goldman and Slattery, *Maintainability – A Major Element of System Effectiveness*, Wiley, New York (1964)

Jensen and Petersen, *Burn In*, Wiley, New York (1982)

Kapur, K. C. and Lamberson, L. R., *Reliability in Engineering Design*, Wiley, New York (1977)

Kivensen, G., *Durability and Reliability in Engineering Design*, Pitman, London (1972)

Moubray, J., *Reliability-centred Maintenance*, Butterworth-Heinemann, Oxford (1997) ISBN 0 7506 3358 1

Myers, G. J., *Software Reliability, Principles and Practice*, Wiley, New York (1976)

O'Connor, P. D. T. O., *Practical Reliability Engineering*, 3rd edn, Wiley, Chichester (1991)

Shooman, M., *Software Engineering – Reliability, Design, Management*, McGraw-Hill, New York (1984)

Snedecor and Cochran, *Statistical Methods*, Iowa State University Press (1967)

Smith, C. O., *Introduction to Reliability in Design*, McGraw-Hill, New York (1976)

Smith, D. J., *Statistics Workshop*, Technis, Tonbridge (1991) ISBN 0 9516562 0 1

Smith, D. J., *Achieving Quality Software*, 3rd edn, Chapman Hall, London (1995) ISBN 0 412 62270X

Smith and Simpson, *Functional Safety, A Straightforward Guide to IEC 61508*, Butterworth-Heinemann, Oxford (2001) ISBN 07506 5270 5

OTHER PUBLICATIONS

Human Reliability Assessors Guide (SRDA-R11), June 1995, UKAEA, Thomson House, Risley, Cheshire WA3 6AT ISBN 085 3564 205

Nomenclature for Hazard and Risk Assessment in the Process Industries, 1985 (reissued 1992), IChemE, 165 - 171 Railway Terrace, Rugby, CV21 3HQ ISBN 0852951841

Tolerability of Risk for Nuclear Power Stations, UK Health and Safety Executive ISBN 0118863681

HSE, 1999, *Reducing Risks Protecting People. A discussion document.* UK Health and Safety Executive.

HAZOP - A Guide to Hazard and Operability Studies, 1977, Chemical Industries Association, Alembic House, 93 Albert Embankment, London SE1 7TU
A Guide to the Control of Industrial Major Accident Hazards Regulations, HMSO, London (1984)
UPM3.1: A Pragmatic Approach to Dependent Failures Assessment for Standard Systems, UK AEA, ISBN 0853564337

STANDARDS AND GUIDELINES

BS 2011 Basic environmental testing procedures
BS 4200 Guide on the reliability of electronic equipment and parts used therein
BS 4778 Glossary of terms used in quality assurance (Section 3.2: 1991 is Reliability)
BS 5760 Reliability of systems, equipment and components
BS 6651: 1990 Code of practice for protection of structures against lightning
UK DEF STD 00-40 Reliability and maintainability
UK DEF STD 00-41 MOD Practices and procedures in reliability and maintainability
UK DEF STD 00-55 The procurement of safety critical software in defence equipment
UK DEF STD 00-56 Hazard analysis and safety classification of the computer and programmable electronic system elements of defence equipment
UK DEF STD 00-58 A Guideline for HAZOP studies on systems which include programmable electronic systems
UK DEF STD 07-55 Environmental testing
US Military Handbook 217E (Notice 1) Reliability Prediction of Electronic Equipment, 1990
US Military Handbook 338 Electronic Reliability Design Handbook
US Military Standard 470 Maintainability programme requirements, 1966
US Military Standard 471A Maintainability/Verification/Demonstration/
Evaluation, 1973
US Military Handbook 472 Maintainability Prediction, 1966
US Military Standard 721B Definitions of Effectiveness Terms for Reliability
US Military Standard 756 Reliability Prediction
US Military Standard 781C Reliability Design Qualification and Production Acceptance Tests
US Military Standard 785A Reliability Programme for Systems and Equipment Development and Production, 1969
US Military Standard 810 Environmental Test Methods
US Military Standard 883 Test Methods and Procedures for Microelectronic Devices
US Military Standard 1629A Procedures for Performing a Failure Mode, Effects and Criticality Analysis
US Military Standard 52779 (AD) Software Quality Assurance Requirements
UK HSE Publication, *Guidance on the Use of Programmable Electronic Systems in Safety Related Applications* (1987)
IGasE Publication SR15, *Programmable Equipment in Safety Related Applications* (third edition 1998) and amendment 2000. ISSN 0367 7850
IGasE Publication SR24, *Risk Assessment Techniques* (1998)
IEE, *Competency Guidelines for Safety Related Systems Practitioners*, 1999, ISBN 085296787X.
IEC Publication 271, *Preliminary List of Basic Terms and Definitions for the Reliability of Electronic Equipment and Components (or parts) used therein*

IEC61508 *Functional Safety: safety related systems* – 7 Parts
IEC International Standard 61511: *Functional Safety – Safety Instrumented Systems for the Process Industry Sector*
IEEE standard 500, Reliability Data for Pumps, Drivers, Valve Actuators and Valves, 1994, Library of Congress 83–082816.
Draft European Standard prEN 51026: *Railway Applications – The Specification and Demonstration of Dependability, Reliability, Maintainability and Safety (RAMS).*
RTCA DO–178B/(EUROCAE ED–12B) – *Software Considerations in Airborne Systems and Equipment Certification.*
UKOOA: *Guidelines for Process Control and Safety Systems on Offshore Installations.*

JOURNALS

Journal of the Safety and Reliability Society (quarterly)
Quality and Reliability International, Wiley (quarterly)
Microelectronics and Reliability (Pergamon Press).
US IEEE Transactions on Reliability.

Appendix 10 Scoring criteria for BETAPLUS common cause model

1 CHECKLIST AND SCORING FOR EQUIPMENT CONTAINING PROGRAMMABLE ELECTRONICS

Score between 0 and 100% of the indicated maximum values.

(1) SEPARATION/SEGREGATION	A Max score	B Max score
Are all signal cables separated at all positions?	15	52
Are the programmable channels on separate printed circuit boards?	85	55
OR are the programmable channels in separate racks?	90	60
OR in separate rooms or buildings?	95	65
MAXIMUM SCORE	110	117

(2) DIVERSITY/REDUNDANCY

	A Max score	B Max score
Do the channels employ diverse technologies?; 1 electronic + 1 mechanical/pneumatic	100	25
OR 1 electronic or CPU + 1 relay based	90	25
OR 1 CPU + 1 electronic hardwired	70	25
OR do identical channels employ enhanced voting? i.e. 'M out of N' where N>M+1	40	25
OR N=M+1	30	20
Were the diverse channels developed from separate requirements from separate people with no communication between them?	20	–
Were the 2 design specifications separately audited against known hazards by separate people and were separate test methods and maintenance applied by separate people?	12	25
MAXIMUM SCORE	132	50

(3) COMPLEXITY/DESIGN/APPLICATION/MATURITY/ EXPERIENCE

	A Max score	B Max score
Does cross-connection between CPUs preclude the exchange of any information other than the diagnostics?	30	–
Is there > 5 years experience of the equipment in the particular environment?	–	10
Is the equipment simple < 5 PCBs per channel? **OR** < 100 lines of code **OR** < 5 ladder logic rungs **OR** < 50 I/O and < 5 safety functions?	–	20
Are I/O protected from over-voltage and over-current and rated > 2:1?	30	–
MAXIMUM SCORE	60	30

(4) ASSESSMENT/ANALYSIS and FEEDBACK of DATA

	A Max score	B Max score
Has a combination of detailed FMEA, Fault Tree analysis and design review established potential CCFs in the electronics?	–	140
Is there documentary evidence that field failures are fully analysed with feedback to design?	–	70
MAXIMUM SCORE	–	210

(5) PROCEDURES/HUMAN INTERFACE

	A Max score	B Max score
Is there a written system of work on site to ensure that failures are investigated and checked in other channels? (including degraded items which have not yet failed)	30	20
Is maintenance of diverse/redundant channels staggered at such an interval as to ensure that any proof-tests and cross-checks operate satisfactorily between the maintenance?	60	–
Do written maintenance procedures ensure that redundant separations as, for example, signal cables are separated from each other and from power cables and must not be re-routed?	15	25
Are modifications forbidden without full design analysis of CCF?	–	20
Is diverse equipment maintained by different staff?	15	20
MAXIMUM SCORE	120	85

(6) COMPETENCE/TRAINING/SAFETY CULTURE

	A Max score	B Max score
Have designers been trained to understand CCF?	–	100
Have installers been trained to understand CCF?	–	50
Have maintainers been trained to understand CCF?	–	60
MAXIMUM SCORE	–	210

(7) ENVIRONMENTAL CONTROL	A Max score	B Max score
Is there limited personnel access?	40	50
Is there appropriate environmental control? (e.g. temperature, humidity)	40	50
MAXIMUM SCORE	80	100

(8) ENVIRONMENTAL TESTING	A Max score	B Max score
Has full EMC immunity or equivalent mechanical testing been conducted on prototypes and production units (using recognized standards)?	–	316
MAXIMUM SCORE	–	316

	A Max score	B Max score
TOTAL MAXIMUM SCORE	**502**	**1118**

2 CHECKLIST AND SCORING FOR NON-PROGRAMMABLE EQUIPMENT

Only the first three categories have different questions as follows:

(1) SEPARATION/SEGREGATION	A Max score	B Max score
Are the sensors or actuators physically separated and at least 1 metre apart?	15	52
If the sensor/actuator has some intermediate electronics or pneumatics, are the channels on separate PCBs and screened?	65	35
OR if the sensor/actuator has some intermediate electronics or pneumatics, are the channels indoors in separate racks or rooms?	95	65
MAXIMUM SCORE	110	117

(2) DIVERSITY/REDUNDANCY	A Max score	B Max score
Do the redundant units employ different technologies?; e.g. 1 electronic or programmable + 1 mechanical/pneumatic	100	25
OR 1 electronic, 1 relay based	90	25
OR 1 PE, 1 electronic hardwired	70	25
OR do the devices employ 'M out of N' voting where; $N>M+1$	40	25
OR $N=M+1$	30	20
Were separate test methods and maintenance applied by separate people?	32	52
MAXIMUM SCORE	132	50

(3) COMPLEXITY/DESIGN/APPLICATION/MATURITY/ EXPERIENCE	A Max score	B Max score
Does cross-connection preclude the exchange of any information other than the diagnostics?	30	–
Is there > 5 years experience of the equipment in the particular environment?	–	10
Is the equipment simple e.g. non-programmable type sensor or single actuator field device?	–	20
Are devices protected from over-voltage and over-current and rated >2:1 or mechanical equivalent?	30	–
MAXIMUM SCORE	60	30

(4) ASSESSMENT/ANALYSIS and FEEDBACK OF DATA
As for Programmable Electronics (see above)

(5) PROCEDURES/HUMAN INTERFACE
As for Programmable Electronics (see above)

(6) COMPETENCE/TRAINING/SAFETY CULTURE
As for Programmable Electronics (see above)

(7) ENVIRONMENTAL CONTROL
As for Programmable Electronics (see above)

(8) ENVIRONMENTAL TESTING
As for Programmable Electronics (see above)

	A Max score	B Max score
TOTAL MAXIMUM RAW SCORE (Both programmable and non-programmable lists)	502	1118

The diagnostic interval is shown for each of the two (programmable and non-programmable) assessment lists. The **(C)** values have been chosen to cover the range 1–3 in order to construct a model which caters for the known range of BETA values.

For Programmable Electronics

	Interval < 1 min	Interval 1–5 mins	Interval 5–10 mins	Interval >10 mins
Diagnostic coverage				
98%	3	2.5	2	1
90%	2.5	2	1.5	1
60%	2	1.5	1	1

For Sensors and Actuators

	Interval < 2 hrs	Interval 2 hrs–2 days	Interval 2 days–1 week	Interval >1 week
Diagnostic coverage				
98%	3	2.5	2	1
90%	2.5	2	1.5	1
60%	2	1.5	1	1

A score of $C > 1$ may only be proposed if the resulting action, initiated by the diagnostics, has the effect of preventing or invalidating the effect of the subsequent CCF failure. For example, in some process industry equipment, even though the first of the CCF failures was diagnosed before the subsequent failure, there would nevertheless be insufficient time to take action to maintain the process. The subsequent (second) CCF failure would thus occur before effective action could be taken. Therefore, in such a case, the diagnostics would not help in defending against CCF and a $C > 1$ score cannot be proposed in the assessment.

AVAILABLE IN SOFTWARE FORM AS BETAPLUS FROM THE AUTHOR AT TECHNIS
01732 352532

Appendix 11 Example of HAZOP

Sour gas consisting mainly of methane (CH_4) but with 2% hydrogen sulphide (H_2S) is routed to an amine absorber section for sweetening. The absorber uses a 25:75% diethanolamine (amine)/water solution to remove the H_2S in the absorber tower. Sweet gas is removed from the tower top and routed to fuel gas. Rich amine is pressurized from the tower bottom under level control and then routed to an amine regeneration unit on another plot. Regenerated amine is returned to the amine absorber section and stored in a low pressure buffer storage tank.

EQUIPMENT DETAILS

Absorber tower operating pressure = 20 bar gauge.

The buffer storage tank is designed for low pressure, with weak seam roof and additional relief provided by a hinged manhole cover.

HAZOP WORKSHEETS

The HAZOP worksheets with this example will demonstrate the HAZOP method for just one node, i.e. the line from the buffer storage tank to the absorber tower.

Nodes that could have been studied in more detail are:

- amine buffer tank
- line to absorber tower from amine buffer tank
- sour gas line to absorber tower
- absorber tower
- sweet gas line out of absorber tower
- rich amine line out of absorber tower,

POTENTIAL CONSEQUENCES

The importance of the consequences identified for a process deviation, and how these are used to judge the adequacy of safeguards, cannot be over emphasized. In this example, the consequences of reverse flow include:

- possible tank damage
- release of a flammable gas near a congested unit which could lead to an explosion
- release of a highly toxic gas.

The latter two consequences alone are deemed sufficient for the matter to be referred back for more consideration. If only the first consequence applied, tank damage could be deemed acceptable if the incident were unlikely, no hazardous substance involved and no personnel would be present. In the common case of a pump tripping and a non-return valve failing, even this may not be deemed acceptable to the HAZOP team if excessive costs from lost production followed from tank damage.

Considerable judgement is called for by the team in making this decision. It is essential that the team be drawn from personnel with sufficient practical knowledge of the process under study.

Although the main action in this example is to consider fitting a slam-shut valve, it could be that an alarm and manual isolation is acceptable. This decision cannot, however, be made without full consideration of the unit manning levels, what duties the operator has that could cause distraction from responding to an alarm, how sufficient will the operator's training be to understand the implications of that alarm, and how far the control panel is from the nearest manual isolation valve.

Figure A11.1 Amine absorber section

Worksheet

Company : Any Town Gas Producers
Facility : Amine Absorber Section
Session : 1 25-07-96
Node : 1 Line from amine tank via pump to absorber tower
Parameter : Flow

Deviation	Causes	Consequences	Safeguards	Recommendations	By
No flow	Amine buffer tank empty	Damage to pump	Level indication	Consider a low level alarm	
		Loss of fresh amine to absorber tower giving H_2S in the sweet gas line	Ditto	Ditto	
	Line frozen	Ditto	Ditto	Check freezing point of water/amine mixture	
	Valve in line shut	Possible damage to line as pump dead heads, i.e. runs against closed discharge line	Operator training	Check line for maximum pump pressure	
More flow	None (fixed by maximum pump discharge)				
Less flow	Line partially plugged or valve partially closed	Possible damage to line as pump dead heads grind against closed discharge line	None	Check freezing point of water/amine mixture and check pipe spec against pump dead head pressure	
Reverse flow	Pump trips	Back flow of 20 bar gas to amine tank	Non-return valve (which may not be reliable in amine service)	In view of the potential consequence of the release and its likelihood, undertake a full study of the hazards involved, and safeguards appropriate to these hazards proposed (possibly installing a chopper valve to cut in and prevent back flow)	
		Resulting in: (1) Possible rupture of tank (2) Major H_2S release to plant causing potential toxic cloud and possible vapour cloud explosion if cloud reaches congested part of the plant	Tank weak seam / None	Ditto / Ditto	
High temperature	Failure of cooling on the amine regeneration unit resulting in hot amine in amine tank	Possibility of poor absorber tower efficiency	Temperature alarm on amine regeneration unit		
Low temperature	Cold conditions	Possible freezing of line	None at present – but see action under 'No flow' to investigate freezing point		
High pressure	Pump dead head	Possibility of overpressure of pipe	None – but see action under 'No flow' to check pipe spec		
	Reverse flow from absorber tower	Ditto	None	In previous action to check pipe spec against pump dead head pressure also include checking spec against operating pressure in absorber tower	
Low pressure	None identified	Not seen as a problem	Line good for vacuum conditions	None	

Appendix 12 HAZID checklist

1 Acceleration/shock	Change in velocity, impact energy of vehicles, components or fluids	1 Structural deformation 2 Breakdown by impact 3 Displacement of parts or piping 4 Seating or unseating valves or electrical contacts 5 Loss of fluid pressure head (cavitation) 6 Pressure surges in fluid systems 7 Disruption of metering equipment
2 Chemical energy	Chemical disassociation or replacement of fuels, oxidizers, explosives, organic materials or components	1 Fire 2 Explosion 3 Non-explosive exothermic reaction 4 Material degradation 5 Toxic gas production 6 Corrosion fraction production 7 Swelling of organic compounds
3 Contamination	Producing or introducing contaminants to surfaces, orifices, filters, etc.	1 Clogging or blocking of components 2 Friction between moving surfaces 3 Deterioration of fluids 4 Degradation of performance sensors or operating components 5 Erosion of lines or components 6 Fracture of lines or components by fast moving large particles
4 Electrical energy	System or component potential energy release or failure. Includes shock both thermal and static	1 Electrocution 2 Involuntary personnel reaction 3 Personnel burns 4 Ignition of combustibles 5 Equipment burnout 6 Inadvertent activation of equipment or ordinance devices 7 Necessary equipment unavailable for functions or caution and warning 8 Release on holding devices

5 Human capability	Human factors including perception, dexterity, life support and error PROBABILITY	1 Personal injury due to: • restricted routes • hazardous location • inadequate visual/audible warnings 2 Equipment damage by improper operation due to: inaccessible control location inadequate control/display identification
6 Human hazards	Conditions that could cause skin abrasions, cuts, bruises, etc.	1 Personal injury due to: • sharp edges/corners • dangerous heights • unguarded floor/wall openings
7 Interface/ interaction	Compatibility between systems/subsystems/ facilities/software	1 Incompatible materials reaction 2 Interfacing reactions 3 Unintended operations caused/prevented by software
8 Kinetic energy	System/component linear or rotary motion	1 linear impact 2 Disintegration of rotating components
9 Material deformation	Degradation of material by corrosion, ageing, embrittlement, oxidation, etc.	1 Change in physical or chemical properties 2 Structural failure 3 Delamination of layered material 4 Electrical short circuiting
10 Mechanical energy	System/component potential energy such as compressed springs	1 Personal injury or equipment damage from energy release
11 Natural environment	Conditions including lightning, wind, projectiles, thermal, pressure, gravity, humidity, etc.	1 Structural damage from wind 2 Electrical discharge 3 Dimension changes from solar heating
12 Pressure	System/component potential energy, including high/low or changing pressure	1 Blast/fragmentation from container overpressure rupture 2 line/hose whipping 3 Container implosion/explosion 4 System leaks 5 Heating/cooling by rapid changes 6 Aeroembolism, bends, choking or shock

13 Radiation	Conditions including electromagnetic, ionizing, thermal or ultraviolet radiation	1 Electronic equipment interference 2 Human tissue damage 3 Charring of organic materials 4 Decomposition of chlorinated hydrocarbons into toxic gases 5 Ozone or nitrogen oxide generation
14 Thermal	System/component potential energy, including high low or changing temperature	1 Ignition of combustibles 2 Ignition of other reactions 3 Distortion of parts 4 Expansion/contraction of solids or fluids 5 Liquid compound stratification 6 Personal injury
15 Toxicants	Adverse human effects of inhalants or ingests	1 Respiratory system damage 2 Blood system damage 3 Body organ damage 4 Skin irritation or damage 5 Nervous system effects
16 Vibration/sound	System/component produced energy	1 Material failure 2 Personal fatigue or injury 3 Pressure/shock wave effects 4 Loosening of parts 5 Chattering of valves or contacts 6 Contamination interface

Index

FARADIP . THREE

FAILURE RATE DATA IN PERSPECTIVE

FAILURE RATES, FAILURE MODES AND A FAILURE MODE AND EFFECT PACKAGE

*Now Issue 4.2 (2001) including latest data sources
- even larger database -*

* Nested menus of failure rates and failure modes from over 30 diverse data sources including military, telecommunications, offshore operations, process and aerospace industries. Based on over 7,000 lines of data.

* Ranges of failure rates spanned by the majority of published sources and an indication of the most common values

* FMEAs for two system failure modes together with parts count and MTBF calculations

* Full editing facilities for globals, files, component entries, etc

>   MICROELECTRONICS (Logic, Linear & Memory)
>   DISCRETE SEMICONDUCTORS, TUBES, LAMPS
>   PASSIVE ELECTRICAL COMPONENTS
>   INSTRUMENTS AND ANALYSERS
>   CONNECTORS, SWITCHES, PCBs, CABLES
>   ELECTROMECHANICAL AND ROTATING MACHINERY
>   POWER SUPPLIES, SOURCES AND TRANSFORMERS
>   MECHANICAL ITEMS (incl PUMPS AND VALVES)
>   PNEUMATIC AND HYDRAULIC EQUIPMENT
>   COMPUTERS, DP AND COMMs
>   ALARMS, FIRE PROTECTION, ARRESTORS, FUSES

Users include British Rail, British Gas, Norwegian SINTEF, Siemens, Veritec, BP, UK MOD, GEC, BNFL, NEI, Gaz de France and Safety Systems companies in the Oil and Gas industry.

AVAILABLE FROM:

26 Orchard Drive,
Tonbridge,
Kent TN10 4LG.
Tel: 01732 352532

£399

COMPARE

Calculating Optimum Maintenance PARamEters

SIGNIFICANT COST SAVINGS CAN BE MADE IN SPARES AND MAINTENANCE RESOURCES. COMPARE PROVIDES:

A RELIABILITY CENTRED MAINTENANCE package which enables OPTIMUM SPARES LEVELS, OPTIMUM PROOF TEST INTERVALS and OPTIMUM REPLACEMENT TIMES to be obtained.

*   Having too many spares carries a cost penalty. Too few spares incurs expensive outage time. Using MARKOV ANALYSIS, COMPARE enables the optimum number of spares to be chosen for any combination of LEAD TIME, REPAIR TIME, FAILURE RATE, NUMBER OF ACTIVE ITEMS and the DEGREE OF REDUNDANCY for that item.

*   Coincident dormant failures, otherwise protected by redundant configurations, will lead to costly outage. Too frequent proof testing, to reveal failed redundant units, will incur a waste of maintenance resource. COMPARE enables the optimum proof test interval to be calculated for a given OUTAGE COST, PROOF TEST COST, DOWN TIME and FAILURE RATE.

*   The cost of an unplanned replacement may be much greater than the planned replacement of a part despite its not having failed. There will be an optimum replacement time for parts having a wearout characteristic. COMPARE enables the Weibull parameters of parts to be assessed from field or test data in order for the optimum time to be calculated.

*   The COMPARE User Manual covers the basic theory behind these techniques. It also includes reliability theory and the application of these techniques.

*   FARADIP.THREE provides a failure rate data benchmark which can be used to review the field data which users apply to COMPARE or as a source of data where generic figures are required.

> FARADIP.THREE and COMPARE, if purchased together, are available at 75% of the price of each (ie £299 each). Existing FARADIP.THREE users will be eligible for a 10% reduction in the price of COMPARE.

AVAILABLE FROM

26 ORCHARD DRIVE,
TONBRIDGE,
KENT TN10 4LG
01732 352532

£399

is a straightforward and fast Fault Tree Analysis package which can be assimilated in one hour. It is used on TECHNIS Reliability Workshops and caters for the majority of fault tree sizes encountered in practice.

It provides:

* A SIMPLE AND LOGICAL TREE INPUT METHOD WHICH ENABLES TREE LOGIC TO BE VERIFIED

* CLEAR OUTPUTS PROVIDING CUT SET RANKING BY UNAVAILABILITY AND BY FAILURE RATE (SIX MEASURES OF CUTSET AND EVENT IMPORTANCE)

* CLEAR GRAPHICS WHICH CAN BE EASILY IMPORTED INTO A WORD PROCESSING PACKAGE (GATE SUPRESSION FACILITY)

* THE TECHNIS BETA+ METHOD FOR TAKING ACCOUNT OF COMMON CAUSE FAILURE

* FACILITIES FOR ONE SHOT PROBABILITY SUCH AS HUMAN ERROR

AVAILABLE FROM

TECHNIS

26 ORCHARD DRIVE,
TONBRIDGE,
KENT TN10 4LG
01732 352532

£975